D0466178

THIRD EDITION

Students and Research

Practical Strategies for Science Classrooms and Competitions

WITHDRAWN

APR 0 7 2020

UNBC Library

Authors in alphabetical order:

Julia H. Cothron
Mathematics and Science Center
Richmond, VA

Ronald N. Giese
The College of William and Mary
Williamsburg, VA

Richard J. Rezba
Virginia Commonwealth University
Richmond, VA

UNIVERSITY OF NORTHERN
BRITISH COLUMBIA
LIBRARY
Prince George, BC

 **KENDALL/HUNT PUBLISHING COMPANY**
4050 Westmark Drive   Dubuque, Iowa 52002

Student activity pages and tables that are clearly
designated, for example, with: © Cothron, Giese, &
Rezba, *Students and Research,* Kendall/Hunt, 2000, are
the only pages in this book that may be reproduced
without written permission.

Copyright © 1989, 1993, 2000 by Kendall/Hunt Publishing Company

ISBN 0-7872-6477-6

All rights reserved. No part of this publication may be reproduced,
stored in a retrieval system, or transmitted, in any form or by any
means, electronic, mechanical, photocopying, recording, or otherwise,
without the prior written permission of the copyright owner.

Printed in the United States of America
10 9 8 7 6 5 4 3

Contents

A Rationale for Student Research

Helping students experience the joys of scientific discovery is one of the delights that makes science teaching a rewarding and satisfying career. The authors believe that much of that joy results from doing science and that the skills of observing, hypothesizing, experimenting, and communicating are the right of every student. An examplary K–12 science program should provide students with opportunities to develop and use these skills. Student investigations, no matter how simple or sophisticated, along with the opportunity for students to share their results take science beyond the memorization of answers to the exploration of questions.

NATIONAL SCIENCE STANDARDS

In this third edition of *Students and Research,* we have included new features that make it easier for classroom teachers to teach students the skills needed to successfully conduct, analyze, and report an experiment. Student experiments are the very kind of inquiry learning described in the National Science Education Standards developed by the National Research Council (NRC) of the National Academy of Science, Project 2061, coordinated by the American Association for the Advancement of Science (AAS), and the Scope, Sequence, and Coordination Project of the National Science Teachers Association (NSTA). The standards, benchmarks, and guidelines of these national reform efforts uniformly emphasize the need to create learning environments that encourage students' understanding of the scientific endeavor and their excitement and enjoyment in its pursuit.

Meeting National Science Standards

National Standards that are specifically met by the suggested instructional activities are noted at the beginning of each chapter. Although the National Science Education Standards (NSES) are organized into six different categories, we have focused our attention on Standard A of the Science Content Standards for grades 5–8 and 9–12. Standard A outlines what students should know about Science as Inquiry and includes skills and abilities such as "Design and conduct a scientific investigation" and "Use appropriate tools and techniques to gather, analyze, and interpret data." We have also included the National Council of Teachers of Mathematics (NCTM) Standards for School Mathematics that are directly related to scientific data display and analysis. These science and mathematics standards are identified in each chapter as (NSES) and (NCTM).

GOAL OF SCIENCE EDUCATION

A guiding principle of the national science education reform effort is that the intellectual traditions of science will pervade the K–12 science curriculum. Those traditions include modes of scientific inquiry, rules of evidence, and ways of formulating research questions that characterize the dominant culture of contemporary science. As we enter the twenty-first century, there is general agreement that the central goal of science education is to prepare a scientifically literate citizenry.

As a nation, we expect our students to leave school with a level of scientific understanding that is essential for carrying out their duties as citizens and being productive members of the American and global economies. To this end, we have reorganized and rewritten the more challenging middle chapters of the book and included new laboratory activities to illustrate the concepts presented in these chapters. A new Chapter 10 has been added for teachers to help students learn new ways to understand and display dispersion and variation in data.

Although there is no agreement on a single definition of scientific literacy, there is consensus that it encompasses three aspects of science—product, process, and habits of mind.

- The *product* part of science is the information base of the life, physical, and earth/space sciences that also includes knowledge of the nature of science, particularly its methods of investigation.
- The *processes* of science are the thinking skills used to solve problems and to conduct inquiries.
- The scientific *habits of mind* are the beliefs and attitudes that characterize the enterprise of science, such as respect for logic and longing to know.

Science Classrooms

The only formal experience most students will have with science is provided by their science teachers. Teaching behaviors and choice of course content influence the way students view science. Students use observations of what occurs in a science classroom to make inferences about the nature of science and to make decisions about continued study in the sciences. Research suggests that when students fail to experience the wonder of scientific investigation and discovery, they avoid future contact with science courses in high school and beyond.

Recognizing how crowded the science curriculum already is, we have included strategies at the end of chapters to help integrate student research skills into the existing science curriculum. Numerous suggestions are provided for modifying current laboratory activities to teach the skills of designing experiments and analyzing data. The features of this new edition will make it easier for you to make your science classroom what you want it to be.

We wish you much success and would appreciate your comments and suggestions.

·········· How To Use This Book ··········

ORGANIZATION OF CONTENTS

This edition of *Students and Research* is presented in four parts.

- **Part One: Basic Principles of Experimental Design and Data Analysis** presents field-tested teaching strategies for developing research skills in students at the elementary, middle, senior high, and introductory college levels. Techniques for helping students design experiments, generate ideas, construct tables and graphs, and write simple reports are given.
- **Part Two: Advanced Principles of Experimental Design and Data Analysis** includes information on how to expand students' research skills such as sophisticated data analysis, experiment design, and reporting. These skills include using library resources, applying descriptive and inferential statistics, designing complex experiments, and writing formal papers.
- **Part Three: Management Strategies for Classroom and Independent Research** gives classroom management strategies to successfully integrate experimental design and data analysis skills into the curriculum and to promote strong school-community ties.
- **Part Four: Strategies for Successful Science Competitions** presents options for maximizing student success and developing positive attitudes about science competitions. Ideas for making effective oral presentations and displays, preparing judges, and selecting competitive events are explored in Part Four.

FEATURES NEW TO THIS EDITION

- **Alignment to National Standards**—standards that are met in each chapter are identified as (NSES) and (NCTM).

- **Technology**—integration of up-to-date technology is an exciting feature new to this edition.
  - **Graphing Calculator**—each investigation includes an optional section on how to use the graphing calculator for the analysis and display of resulting data. User-friendly, step-by-step directions for using graphing calculators are given in *Appendix A: Using Technology.*
  - **Internet**—hundreds of helpful web site addresses are included throughout the text. Every effort has been made to provide up-to-date addresses of these sites, but they may change.
- **Assessments**—because evaluation often sends a message of what is considered important, the focus has continued on multiple forms of assessment. Checklists, rating sheets, and numerous suggestions for test items are included in individual chapters and in Chapters 15 and 16. *Appendix B: Practice Problems Answer Key* provides the answers to suggested performance tasks and the test items that are given throughout the book.
  - Additional assessment items are provided in the new edition of *Science Experiments and Projects for Students,* the student version of *Students and Research.*
- **Chapter objectives**—at the beginning of each chapter objectives are given for the teacher and the student.
  - The symbol *T* is used for teacher objectives.
  - The symbol *S* is used for students' objectives when appropriate to the chapter.
- **Chapters**—have been reorganized and rewritten with new laboratory activities. A new *Chapter 10 Displaying Dispersion/Variation in Data* has been added to help students learn new ways to understand this concept.

- **Strategies at the end of chapters**—given to help integrate student research skills into the existing science curriculum.
- **Modifying current laboratory activities**—numerous suggestions are provided in each chapter for modifying current laboratory activities in existing curriculum to teach new concepts and skills.
- **Black line duplicating masters**—directly related to the content and skills presented in chapters these include—
  > activities
  > investigations
  > information sheets
  > performance assessments

- *Appendix C: Experimenting Safely*—has been added to emphasize the absolute importance of safety in science.
- **Glossary**—the extensive glossary puts all of the important concepts and terms in one multi-page list.
- **Index**—provides page references to terms and concepts.

About the Authors

- **Julia H. Cothron, Ed.D.** has worked with middle and high school teachers to develop effective strategies for developing students' research skills and has served as a mentor to hundreds of students. From 1980 to 1991 she supervised the Hanover County Public Schools (Virginia) science program, which has earned a reputation for excellence in student research. Currently, she is Executive Director for the Mathematics & Science Center, a consortium of six school divisions, located in Richmond, VA.

- **Ronald N. Giese, Ed.D.** is a professor of science education at The College of William and Mary; he has worked with both preservice and inservice teachers to develop strategies for generating research topics and to implement science fairs that maximize student learning. Dr. Giese has served as a consultant to Scholastic *Science World,* to the Naturalist Center at the National Museum of Natural History, Smithsonian Institution, and to numerous school systems, museums, and science curricular projects.

- **Richard J. Rezba, Ph.D.** is a professor of science education at Virginia Commonwealth University where he works with both elementary and secondary teachers to develop instructional strategies in science that are both challenging and fun. His research interests include parental involvement, student experimentation, and assessment. Dr. Rezba directs several projects that involve the infusion of various forms of instructional technology into the teaching and learning of science.

The authors are successful teachers who have written extensively and have conducted numerous workshops and courses at state and national levels for K–12 teachers and university faculty. They are the joint recipients of a Distinguished Achievement Award for Excellence in Educational Journalism from the Educational Press Association of America.

Acknowledgments

Although it would be impossible to acknowledge everyone who assisted us with the original manuscript and the editions that followed, we would like to express our gratitude to the people who offered us valuable advice, practical suggestions, technical assistance, emotional support, and encouragement throughout the preparation of the manuscript.

Our thanks to teachers in Virginia, West Virginia, North Carolina, Georgia, Florida, Louisiana, Texas, Washington, Missouri, Pennsylvania, Massachusetts, New Hampshire, Kentucky, Michigan, Tennessee, and Washington, D.C. who participated in staff development workshops, and who shared their knowledge with us.

Jeannette Bishop, Nancy Cozart, Bruce Maxwell, Arthur Schwieder, Eleanor Tenney, and Wilton Tenney for photographs.

*Our spouses—Sam, Bobbie, and Carol—whose support and encouragement made **Students and Research** possible.*

PRIOR PUBLICATIONS

Certain chapters of this book are related to articles published in science education journals. (*Indicates equal co-authorship, Authors Alphabetically Listed)

- Rezba, R.J., Giese, R.N., & Cothron, J.H. (Jan. 1998). Graphs, Let the data do the talking. *Science Scope*. (Chapter 5)
- *Cothron, J.H., Giese, R.N., & Rezba, R.J. (1995). Designer airplanes. In R. Moore (ed.) *Biology on a shoestring*. Reston, VA: The National Association of Biology Teachers Press. (Chapter 1)
- Cothron, J.H., Giese, R.N., & Rezba, R.J. (1994). Simple principles of data analysis. In R. Moore (ed.) *Biology labs that work: The best of how-to-do-its*. Reston, VA: The National Association of Biology Teachers Press. (Chapter 8)
- Rezba, R.J., Giese, R.N., & Cothron, J.H. (1993). Dining on data: Making a serving table. *Science Scope, 17*(3), 26–29. (Chapter 5)
- *Cothron, J.H., Giese, R.N., & Rezba, R.J. (1993). How to launch a science experiment (7–11) (Chapter 1); Bursting with project ideas (12–13) (Chapter 3); Ready, aim, splat! (14–16) (Chapter 5). *Science World, 50*(2).
- *Cothron, J.H., Giese, R.N., & Rezba, R.J. (1992). Fold it (8–16) (Chapter 1); Stuck for a project idea? (17–19) (Chapter 3); Spring into action (20–22) (Chapter 4); All wrapped up (23–26) (Chapter 6); *Science World, 49*(2).
- Rezba, R.J., Giese, R.N., & Cothron, J.H. (1992). Traditional labs plus new questions equal improved student performance. *Science Scope, 15*(5), 39–44. (Chapters 1, 2, & 3)
- Giese, R.N., Cothron, J.H., & Rezba, R.J. (1991). Take the search out of research. *The Science Teacher, 59*(1), 32–37. (Chapter 7)
- *Cothron, J.H., Giese, R.N., & Rezba, R.J. (1991, 1992). What is an experiment? (2–5) (Chapter 1); No pain, no gain (6–7) (Chapter 2); Step by step procedure (8–9) (Chapter 4); Shine your design (10–11) (Chapter 3); Science safe (12) (Appendix); Serving up your results (13–15) (Chapter 5); Write it right (16–19) (Chapter 6); *Science World, 48* (7 & 8).

- *Cothron, J.H., Giese, R.N., & Rezba, R.J. (1990). Come fly with us (2–5) (Chapters 1 & 2); Four easy pieces (6–7) (Chapter 3); *Science World, 47*(7).
- Cothron, J.H., Giese, R.N., & Rezba, R.J. (1990). Basic concepts of experimental design. *Science Scope, 13*(5), 28–32. (Chapter 1)
- Cothron, J.H., Giese, R.N., & Rezba, R.J. (1989). Simple principles of data analysis. *The American Biology Teacher, 51*(7), 426–428. (Chapter 8)
- Cothron, J.H., Rezba, R.J., & Giese, R.N. (1989). What to keep in mind during experimental design. *The Science Teacher, 56*(8), 33–37. (Chapter 2)
- Cothron, J.H., Rezba, R.J., & Giese, R.N. (1989). Writing results and conclusions. *The American Biology Teacher, 51*(4), 239–242. (Chapters 8 & 9)
- Giese, R.N., Cothron, J.H., & Rezba, R.J. (1989). Procedure writing—But this ain't no English class. *Science Activities, 26*(1), 24–28. (Chapter 4)
- Giese, R.N., Rezba, R.J., & Cothron, J.R. (1989). An open letter to science fair judges: Focus on projects and presenters. *Science Scope, 13*(2), 38–41. (Chapter 19)

Other books by authors:
- *Cothron, J.H., Giese, R.N., & Rezba, R.J. (1996). *Science experiments and projects for students*, Dubuque, IA: Kendall/Hunt Publishing Company
- *Cothron, J.H., Giese, R.N., & Rezba, R.J. (1996). *Science experiments by the hundreds: Experimenting at home and school*, Dubuque, IA: Kendall/Hunt Publishing Company
- *Cothron, J.H., Giese, R.N., & Rezba, R.J. (1996). *Science experiments by the hundreds: Experimenting at home and school (Teachers Guide)*, Dubuque, IA: Kendall/Hunt Publishing Company

PART ONE

Basic Principles of Experimental Design and Data Analysis

✓ Use strategies for presenting inquiry skills to students at the elementary/middle/senior high/introductory college levels to:

- design experiments
- overcome design flaws
- write procedures
- construct tables and graphs
- write simple reports
- generate ideas

Developing Basic Concepts

Objectives

- Identify and define the basic concepts of experimental design.

  hypothesis   independent variable
  dependent variable   constants
  control group   repeated trials

- Modify current lab activities to teach the basic concepts of experimental design.

National Standards Connections
- Identify questions that can be answered through scientific investigation (NSES).
- Systematically collect, organize, and describe data (NCTM).

Entries in local science fairs include models, posters, demonstrations, written reports from encyclopedias, and true scientific experiments. Because guiding student research is a difficult and time consuming task, true scientific experiments may be in the minority. Large class size, insufficient student motivation, inappropriate parental support, and differing student abilities make the task even more challenging. Furthermore, few science textbooks or college courses provide teachers with effective instructional strategies or models for teaching students to design experiments.

A common, but often ineffective, method for introducing experimental design is teaching the steps of the scientific method. Typically, five or six major steps are listed, defined, and illustrated with a model experiment. Students are then directed to find a project topic, select a specific problem, and design an experiment. This approach has many problems. First, the scientific method describes the way experiments are reported, not how they are generated. Second, the instructional sequence does not sufficiently develop the basic concepts of experimental design: hypothesis, independent and dependent variables, constants, control, and repeated trials. Teaching the scientific method in the fall and displaying science projects in the spring further compounds the problem. Students may see little relationship between the two events, as evidenced by many poorly designed experiments. Valuable opportunities to develop and reinforce experimental design concepts through classroom experiments and research discussions are also lost.

An effective method for teaching students to design experiments begins with concrete investigations of phenomena that enable students

to quickly manipulate materials and see the results. By observing these interactions, students develop an intuitive understanding of experimental design that can be crystallized through effective teacher questioning and class discussion. Formal concepts and terms that emerge from these discussions hold meaning for students because they are based on real phenomena. Retention is further enhanced by continuous application of the concepts to textbook and laboratory activities. In this chapter, two examples of this effective technique are illustrated: a **qualitative approach** with paper airplanes and a **quantitative approach** with dissolving chemicals.

DESIGNER PLANES

Distribute sheets of paper. Ask the students to make a paper airplane and write their name on it. You should make an airplane as well. Students may work individually or in groups of two to five. Have the students form a line against a wall. Join them with your plane. "Ready, set, . . . hold it!" Ask the students, "How will we know which plane is best?" Probe for five or six possible criteria, such as the longest distance, the straightest flight, and the amount of flight time. Ask, "Which criterion would be the most difficult to measure?" "Which would be the least difficult to measure?" "Which is the best to use?" and "Why?" End the discussion by selecting one of the criteria. After flying the planes, give the students several minutes to modify their planes as they wish for another flight. Supply materials such as scissors, paper clips, and tape. Do not change your own plane! Have the students make a **hypothesis** or *educated guess* about the effect of the change they made, for example, "**If I add

weight to the wings, then** the plane will fly straighter." Form a student line again. This time ask students whether throwing their planes only once is the fairest procedure for determining which plane is the best. Why not? Encourage the students to suggest that several airplane flights or **repeated trials** would be a more reliable test. Through questions and suggestions, lead students to the concept that repeated trials reduce the effects of chance errors on the overall results (see Investigation 1.1, *Designer Planes*).

Leading Questions

* How did you *act* on your plane?
* What did you *purposely change* about your plane?
* How did you determine your plane's *response*?
* What *remained the same* about your plane?

Write these leading questions on the chalkboard. Be sure the students understand that they acted on their planes by adding paper clips, refolding, and so on. Through these actions, they purposely changed something about the plane such as its weight or wing design. Discuss synonyms for change such as *modified, altered,* and *varied.* Give students several minutes to formulate answers to the questions. On the board list specific student actions, changes made on purpose, and the planes' responses. Also include items that remained the same during the investigation.

Introduce the term **variable** to describe each factor that changed in the experiment. Distinguish between the variables that were purposely changed or manipulated, the **independent variables,** and the variables that responded, the

Action	Purposely changed	Response to change	Remained the same
	Independent variable	*Dependent variable*	*Constant*
Refolded	Wing shape	Flight time	Size of paper
Added clips	Center of gravity	Total distance	Texture of paper
Added tape	Weight of plane	Straight travel	Weight of paper

INVESTIGATION 1.1 • Designer Planes

Materials

- Paper
- Scissors
- Tape
- Paper clips
- Safety goggles

Safety

- Wear goggles.
- Do not throw planes at fellow students.
- Handle sharp objects safely.

Procedure

Part I

1. Make a paper airplane. Write your name on it.
2. Follow your teacher's directions for flying the plane.
3. Observe the plane's flight.

Part II

4. Use the materials to modify your plane.
5. Make a **hypothesis** about the effect of the change.
6. Follow your teacher's directions for flying the plane.
7. Observe the plane's flight.

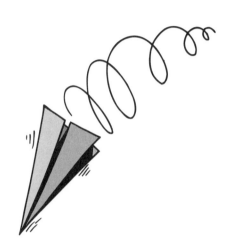

Thought Questions

1. How did you **act** upon your plane?
2. What did you **purposely change** about your plane?
3. How did you determine your plane's **response?**
4. What **remained the same** about your plane?

Class Data Table

Question 1	Question 2	Question 3	Question 4
Action	**Purposely Changed**	**Response to Change**	**Remained the Same**

© 2000 by Kendall/Hunt Publishing Company, Cothron, Giese, & Rezba, *Students and Research*.

dependent variables. Next, list all the factors that remained the same for each of the original planes, e.g., size, weight, and texture of paper. Label the factors that remained the same as **constants.** Add the terms to the class data table as you introduce them (see Investigation 1.1).

Suggest to students that unnoticed changes in the room's air flow from heating units, doors, or windows may have affected their plane's flight. Ask how they could be sure that the change they purposely made in their plane really affected its flight, rather than some unknown factor. Lead students to the conclusion that one way would be to fly at least one unmodified airplane in each trial to see whether its flight remained the same. Tell them about your unchanged plane and how it flew the same in both flights. Introduce the term **control** to describe the plane that was unchanged and that was used to detect and measure any effects of unknown factors.

At this point, students should be able to identify the hypothesis, independent variable, dependent variable, constants, control, and repeated trials in the investigation. For students to apply the concepts, however, numerous experiences will be necessary. Provide additional practice through modified textbook experiments or supplemental investigations.

HOT SOLUTIONS

Although many students realize that solids dissolve better in hot water, few realize that the dissolving process may affect the water temperature. Substances that raise the water temperature include calcium chloride ($CaCl_2$), anhydrous magnesium sulfate ($MgSO_4$), and anhydrous copper sulfate ($CuSO_4$). Other chemicals like ammonium chloride (NH_4Cl) and sodium hyposulfite or photographer's hypo ($NaHSO_3$) will decrease water temperature.

Because calcium chloride prevents icing of walkways and roads, it is available through hardware and agricultural suppliers, as well as chemical vendors. Establish a mindset for Investigation 1.2, *Hot Solutions,* by demonstrating the dissolving of calcium chloride in water. Ask students to describe the process. Add your observation that the beaker warmed as the calcium chloride dissolved. Ask, "What would happen if more calcium chloride were added to the water?" Formulate a **hypothesis** or educated guess that describes the effect of adding more calcium chloride to the water. Frequently, a hypothesis is stated as an **if . . . then . . .** format, for example, **If** more scoops of calcium chloride are added to water, **then** the temperature of the water will increase. Describe a list of materials and a procedure that students can use to test the hypothesis (see Investigation 1.2, *Hot Solutions*). To save time, give each pair of students a specific amount of calcium chloride, such as 0, 1, 2, or 3 scoops. Quick approximate measurements are adequate. Later, the importance of precise measurements can be discussed. Be sure to emphasize appropriate safety precautions including wearing safety goggles, washing hands, and disposing of chemicals in properly marked containers. Allow students approximately ten minutes to collect and record data in a class data table.

Representative student data for *Hot Solutions* are provided in Table 1.1 Class Data from *Hot Solutions.* Given the normal class size, three measurements for each amount of the chemical can be obtained easily. Using this data, the concept of **averaging** can be taught or reviewed. Emphasize that each measurement made by a group constitutes a separate trial. Ask, "Do you have more confidence in one measurement or the average of several measurements? Why?" Help students understand that **repeated trials** increase confidence by reducing the effects of chance or random errors that may occur in a single trial. Ask the students, "What varied or changed in the experiment?" Introduce the term **variable** to describe all factors that could change in an experiment. Distinguish between the variable that they purposely changed or manipulated, the amount of chemical, and the variable that responded, the temperature of water. Label the former as the **independent variable** and the latter as the **dependent variable.** Some textbooks use the synonyms, **manipulated** and **responding variables.** For very young students, some teachers initially call them the **you change it variable** and the **it changed variable.** Ask, "What factors remained the same or unchanged in the experiment?", for example, amount of water (75 ml), time to dissolve (2 minutes), stirring. Label all the factors that remained the same as **constants.** Next, inquire why measurements were made on water

INVESTIGATION 1.2 • Hot Solutions

Materials
- 150 ml beaker or plastic cup
- Thermometer (°C)
- Scoop or plastic spoon
- Safety goggles
- Chemical (*calcium chloride*)
- Clock
- Water
- Graduated cylinder

Safety
- Wear goggles.
- Wash hands.
- Dispose of chemicals in marked containers.

Procedure
1. Place 75 ml (about 1/3 cup) of water in a beaker or cup.
2. Record the initial temperature of the water (°C) **in the margin**.
3. Measure the designated amount of the calcium chloride, such as 0, 1, 2, 3 scoops or spoonfuls.
4. Add and stir the designated amount of the chemical for 2 minutes.
5. Record the temperature (°C) of the water at the end of 2 minutes **in the margin**.
6. Calculate the temperature change of the water.
7. Record data on the class data table.

Record & Calculate

°C End Temp
– °C Initial Temp
°C Change in Temp

°C End Temp
– °C Initial Temp
°C Change in Temp

°C End Temp
– °C Initial Temp
°C Change in Temp

°C End Temp
– °C Initial Temp
°C Change in Temp

°C End Temp
– °C Initial Temp
°C Change in Temp

°C End Temp
– °C Initial Temp
°C Change in Temp

°C End Temp
– °C Initial Temp
°C Change in Temp

°C End Temp
– °C Initial Temp
°C Change in Temp

°C End Temp
– °C Initial Temp
°C Change in Temp

°C End Temp
– °C Initial Temp
°C Change in Temp

°C End Temp
– °C Initial Temp
°C Change in Temp

°C End Temp
– °C Initial Temp
°C Change in Temp

| Amount of chemical (scoops) | Change in temperature (°C) | | | Average change in temperature (°C) |
| | Trials | | | |
	1	2	3	
0				
1				
2				
3				

How could you improve this experiment?

USING TECHNOLOGY ·
1. In the **STAT** mode of your calculator, enter the number of scoops in List 1 and the values for mean change in temperature in List 2. (See Appendix A, *Using Technology*, for additional help in using the graphing calculator).
2. In setting up your graph, select scatter plot as your graph type and List 1 for your x values and List 2 for your y values. Graph the data.
3. Press **Trace** and use the arrow keys to highlight each x value (number of scoops) and to see the corresponding y values (temperature).

© 2000 by Kendall/Hunt Publishing Company, Cothron, Giese, & Rezba, *Students and Research*.

TABLE 1.1   Class Data from Hot Solutions

Amount of chemical (scoops)	Change in temperature (°C)			Average change in temperature (°C)
	Trials			
	1	2	3	
0	0	1	1	0.7
1	8	9	7	8.0
2	15	11	14	13.3
3	20	17	17	18.0

containing zero scoops of the chemical. Introduce the concept, **control,** to describe the water samples that received no treatment and that were used to detect or measure the influence of unanticipated factors, such as temperature gain from the room or from stirring. The control is the standard for comparing experimental effects. In some experiments, the control is the group that receives no treatment, whereas in other experiments, the control may be an arbitrarily selected standard, for example, dissolving rate at room temperature or durability of pure wool carpet. Scientists often refer to constants as controlled variables. Because students who are just learning the concepts of variables and controls find the term controlled variables confusing, the authors prefer to reserve the term *control* to refer to the group that is used as a standard for comparison.

MODIFYING YOUR LAB ACTIVITIES

Nothing comes free. The simple modifications suggested here and in similar sections in other chapters require a little time. The reasons these modifications are necessary in the first place is because of text authors' well-intentioned concern for conserving class time. But saving time for what? Perhaps, the reallocation of some instructional time to the essence of science, science experimentation, is long overdue.

Although many textbook activities, such as Modified Example 1.1, *A Working Heart,* involve

the concepts of independent and dependent variables, constants, and sometimes even controls and repeated trials, these concepts are rarely explicit or emphasized. For most laboratory activities that involve manipulating a variable and measuring a response, it is easy to make simple modifications to introduce and practice these basic concepts of experimental design. Asking students questions about the investigation is among the simplest ways to modify or extend an existing activity. In Modified Example 1.1, *A Working Heart*, students could be asked to identify the variable that is purposely changed (temperature of the water) and the variable that responds to that change (the heartbeat of the Daphnia). What was once implicit, now becomes explicit. These variables can be labeled later as the independent and dependent variables, or the manipulated and responding variables depending on your terminology preference.

To develop the importance of keeping all other potential variables constant, students should be asked to identify those factors of an investigation that are kept the same (kind of Daphnia, amount of water, and the procedures followed) and to label these factors as constants. In lab activities similar in format to Modified Example 1.1, *A Working Heart* where the control is neither emphasized nor obvious, students should be asked to identify the part of the experiment that serves as a standard for comparison. In many experiments the control is the group that receives no treatment, but in this Daphnia investigation, the control probably would be designated as the average heartbeat of Daphnia in the temperature of water closest to their normal environment, 20°C. The heartbeat at temperatures higher and lower can then be compared to this designated group that serves as the control.

To increase confidence in a conclusion, an experiment is repeated a number of times to check for consistency of results and to reduce the effects of random errors. In the experiment with Daphnia, students should be asked to repeat the procedure with fresh Daphnia at least five times. Because of restrictions in class time, most lab activities portray an unrealistic picture of science. Repeated trials, however, are an important part of science and students should have some experience with them whenever possible. Because time is not always available students at least should be asked how they could increase the reliability of their results. An alternate procedure for repeated trials, which also conserves time, is to pool the data collected by each set of lab partners in the class. In the Daphnia experiment, for example, several groups could be directed to collect data at each designated temperature. These data can be pooled and then viewed as repeated trials for each temperature value. A table for recording class data can be drawn on the board, and as students complete the investigation, they can record their data on the board.

Note how Modified Example 1.1, *A Working Heart* was modified for developing basic concepts of experimental design by the simple addition of a few questions. Examine your lab activities and identify those that could be similarly modified to help students learn and practice the basic concepts of experimental design.

EVALUATING AND REINFORCING SKILLS

Most students will not master the basic concepts of experimental design from one or two investigations. Repeated experiences with varied phenomena over time are necessary. Because of time constraints, these concepts must be reinforced throughout the curriculum. Unfortunately, many science programs do not explicitly develop the concepts of variables, constants, control, repeated trials, or hypothesis. With minor modifications of laboratory activities, however, the concepts can be integrated easily into the curriculum.

Ideas for assessing student achievement with paper-pencil tests are provided in Chapter 15. Questions require students to identify the variables in an experiment, and match terms and definitions.

REFERENCES

National Science Teachers Association. (Yearly). *NSTA supplement of science education suppliers.* Washington, DC: National Science Teachers Association.

Related Web Sites

http://pointer.wphs.K12.va.us/ 118sci.htm (teacher adaptation/use components of an experiment)

http://www.isd77.K12.mn.us/resources/ cf/SciProjIntro.html (Elementary Level)

http://www.isd77.k12.mn.us/ resources/cf/SciProjInter. html (discussion of repeated trials, random errors, and systematic errors)

MODIFIED EXAMPLE 1.1 • A Working Heart

Purpose: To determine the effects of temperature on the heartbeat of Daphnia.

Materials: Daphnia
Microscope
Water at various temperatures
Glass slide
Petroleum jelly
Toothpick

Beaker (250 ml)
Test tube
Graduated cylinder
Foam cup
Medicine dropper

Procedure

Part I

At the central supply area, place 150 ml of water at the temperature assigned to your laboratory group into your foam cup (0°, 10°, 20°, 30°, and 40°C). Place the cup in a 250 ml beaker for greater stability. Put a test tube of water containing several Daphnia into your cup of water, and wait 10 minutes before proceeding. Place a clean glass slide in the cup as well so that the slide will be the same temperature as the Daphnia.

Part II

Acting quickly, prepare the slide for the Daphnia by drying it; using a toothpick, place a small amount of petroleum jelly near the center of the slide. Remove a Daphnia from the test tube using a medicine dropper and place it on the jelly. Quickly place the slide on the stage of a microscope and focus. Count the number of heartbeats in 15 seconds, multiply by four, and record your data in the data table on the chalkboard. Copy the graph below and use the class averages to plot the rate of heartbeat and temperature.

Analysis

1. How does temperature affect the Daphnia's heartbeat?
2. At what temperature is the heartbeat the fastest?
3. Using your graph, what do you think the heartbeat would be at 25°C; at 45°C?

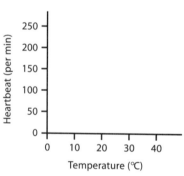

Modifications for Teaching Basic Concepts of Experimental Design

Ask:

1. What variable was purposely changed?
2. What variable responded?
3. Is there a control in the experiment? Justify your answer.
4. What factors remained the same in the experiment?
5. How could you improve the experiment?

Practice

In the "Floor Wax Test" scenario below, identify the following components of an experiment:

1. independent variable
2. dependent variable
3. constants
4. repeated trials
5. control

Use the scenario below to write a title and a hypothesis using the following formats:

6. Title: The Effect of the *(changes in the independent variable)* on the *(dependent variable)*.

7. Hypothesis: If the *(independent variable—describe how it will be changed),* then the *(dependent variable—describe the effect).*

A shopping mall wanted to determine whether the more expensive "Tough Stuff" floor wax was better than the cheaper "Steel Seal" floor wax at protecting its floor tiles against scratches. One liter of each grade of floor wax was applied to each of 5 test sections of the main hall of the mall. The test sections were all the same size and were covered with the same kind of tiles. Five (5) other test sections received no wax. After 3 weeks, the number of scratches in each of the test sections was counted.

Applying Basic Concepts

© 1998 PhotoDisc, Inc.

Objectives

- Define and identify the major concepts of experimental design in a scenario of an experiment.

  hypothesis   independent variable
  dependent variable   constants
  control   repeated trials

- Use the concepts of experimental design to suggest improvements in experiments.
- Construct a simple diagram to communicate the major components of an experiment.
- Use scenarios of experiments to teach the basic concepts of experimental design and to develop improved designs.
- Generate scenarios of experiments that illustrate typical design flaws.

National Standards Connections

- Identify questions and concepts that guide scientific investigation (NSES).

Lists can be useful tools of analysis, but diagrams are frequently much more powerful tools. Students could analyze an experiment, for example, by listing its separate components. However, making a simple diagram of the same experiment that communicates its title, the hypothesis, the independent variable, the dependent variable, the constants, the control, and the number of repeated trials is a more effective way to quickly visualize the design of an experiment. Figure 2.1 Experimental Design Diagram sum- marizes Investigation 1.2, *Hot Solutions* from Chapter 1.

To construct an experimental design diagram for any experiment with one independent variable, follow these steps:

1. Write a **title: The Effect of** the Independent Variable **on** the Dependent Variable.
2. State a **hypothesis: If** the (independent variable) is (describe how you changed it), **then** the (dependent variable) will (describe the effect).

Title: The Effect of Various Amounts of Calcium Chloride on the Temperature of Water
Hypothesis: If more scoops of calcium chloride are added to water, then the temperature of the water will increase.

IV: Amount of calcium chloride (scoops)				←	Independent Variable
0 scoop (control)	1 scoop	2 scoops	3 scoops	←	Levels of Independent Variable Including the Control
3 trials	3 trials	3 trials	3 trials	←	Repeated Trials Number of times each of the levels of IV was tested

DV: Temperature of water   ← Dependent Variable
C: Same amount of water (75 ml)   ← Constants
       Same time to dissolve (2 min)
       Constant stirring

Figure 2.1 Experimental Design Diagram.

3.	Write your **independent variable (IV:).**
4.	Divide the bottom two rows into columns; one column for each **level** of the independent variable. Write the levels of the IV in the columns. If one of those levels is the control for the experiment, put the word **(control)** under that level.
5.	In each column write the number of **repeated trials** conducted for each level of the independent variable.

6. Put your **dependent variable (DV:)** below the rectangle.
7. Write a list of the **constants (C:).**

CHECKING THE DESIGN

Because each part of an experiment has its place in an experimental design diagram, students can spot missing or weak parts quickly and easily. Using the checklist in Table 2.1 *Checking The Experimental Design*, ask students to begin at the top of the experimental design diagram and look for ways to improve the experimental design.

Using the checklist questions to critique the experimental design diagram for the *Hot Solu-*
tions experiment (Figure 2.1) produced the following results:

Q1. Does the title identify both the independent and dependent variables?
The original title, Hot Solutions, *does not identify the independent and dependent variables. This title does,* The Effect of Various Amounts of Calcium Chloride on the Temperature of Water.

Q2. Does the hypothesis clearly state how you think changing the independent variable will affect the dependent variable?
The hypothesis is fine.

Q3. Is there an independent variable?
The amount of calcium chloride is a single, well-defined independent variable.

Q4. Are the levels of the independent variable clearly stated? Are there enough levels of the independent variable tested?
Yes, the levels are sufficient and clearly stated.

Q5. Is there a control? Is it clearly stated?
There is a control, 0 scoops of calcium chloride.

Q6. Are there repeated trials? Are there enough of them?
Three trials are ok, but 5 or more would have been better. A common question is,

TABLE 2.1   Checking the Experimental Design

Checklist Question Number	Questions
☐ 1.	Does the title clearly identify both the independent and dependent variables?
☐ 2.	Does the hypothesis clearly state how you think changing the independent variable will affect the dependent variable?
☐ 3.	Is there just one independent variable? Is it well defined?
☐ 4.	Are the levels of the independent variable clearly stated? Are there enough levels of the independent variable tested? Are there too many?
☐ 5.	Is there a control? Is it clearly stated?
☐ 6.	Are there repeated trials? Are there enough of them?
☐ 7.	Is the dependent variable clearly identified and stated?
☐ 8.	Is the dependent variable operationally defined? Operationally defined means that the investigator very clearly stated how the response would be measured or described?
☐ 9.	Are the constants clearly identified and described? Are there any other?
☐ 10.	Did the experimental design diagram include all the parts listed in Figure 2.1? Were all the parts placed in the proper place?
☐ 11.	Was the experiment creative? Was it an appropriate level of complexity?

"How many trials are enough?" The answer depends on the experiment. There are few differences among granules of calcium chloride or among paper airplanes that are constructed the same way. When you investigate such non-living things, you tend to get very similar results. For this reason, you can use a smaller number of trials, such as 5 to 10. However, there are many differences among even similar looking groups of plants and animals. Because of the many differences in organisms, you tend to get a greater variety of results; therefore, more trials are needed. With human studies, such as studies on taste preferences, even more trials are necessary. With organisms you should generally conduct as many trials as time, money, and space will allow.

Q7. Is the dependent variable clearly identified and stated?
The temperature of the water is not the dependent variable. The change in the temperature of the water is.

Q8. Is the dependent variable operationally defined? Operationally defined means that the investigator very clearly stated how the response would be measured or described.

No, the experimental design diagram did not indicate that the water temperature was measured in degrees Celsius. To improve the diagram, add the symbol (0 °C) to show how the change in water temperature was measured.

*For the dependent variable, you can use either quantitative or qualitative data. **Quantitative** observations include measurements with standard scales, such as degrees Celsius. **Qualitative** observations include verbal descriptions or measurements with non-standard scales. For example, you could describe the cloudiness or color of the solution. You will learn more about these types of dependent variables in Chapter 8, Analyzing Experimental Data.*

Q9. Are the constants clearly identified and described? Are there any others?
Some constants are clearly identified, such as the amount of water (75 ml) and the time to dissolve (2 min). Other constants are unclear. For example, what is the scoop size? Is it a level or heaping scoop? It would be better to use a balance to measure the calcium chloride in grams. What about the initial temperature of the water for each trial? Was

it always the same? If not, this could be a hidden variable affecting the results. What is meant by constant stirring? How could you be sure the stirring was the same?

Other unlisted factors may also have affected the results of the experiment. For example, what type of container was used? Were the containers good or poor insulators? How might the type of container affect the results? What other factors can you identify that might have affected the experimental results?

Q10. Did the experimental design diagram include all the parts listed in Figure 2.1? Were all the parts placed in the proper place.
Yes, all the parts were included and properly placed.

Q11. Was the experiment creative? Was it at an appropriate level of complexity?
Deciding upon creativity and appropriateness are "judgement calls" on your part. For an experiment to be appropriate, it should address a question whose answer is unknown to the investigator. Determining the effect of different colors of ground covers on plant growth is an appropriate experiment, while determining the effect of light versus dark on plant growth is not. Almost everyone knows that plants will die without light. In determining appropriateness, you will also need to evaluate your own background knowledge. What would have been an appropriate experiment for you in sixth grade is not an appropriate experiment for your junior year.*

Have students use the checklist in Table 2.2 *Checklist for Evaluating Experimental Design* to evaluate the experimental design diagrams they construct for experiments. Students can ask a friend or family member to check the diagram as well and make further improvements.

PRACTICING SKILLS

Knowledge of the basic concepts of experimental design and the ability to apply these concepts represent two different levels of learning. Students may define the terms or identify examples

TABLE 2.2   Checklist for Evaluating Experimental Design

Criteria/Value (100 points)	Self	Peer/Family	Teacher
Title (5)			
Hypothesis (5)			
Independent variable (10)			
Levels of independent variable (10)			
Control (10)			
Repeated trials (10)			
Dependent variable (10)			
Operational definition of dependent variable (10)			
Constants (15)			
Experimental design diagram (10)			
Creativity/Complexity (5)			

in a classroom investigation yet fail to apply the concepts when designing an experiment. Remember the student who defined repeated trials in class but designed an experiment with only one trial, or who identified the control group in a textbook activity but designed an experiment without a control. Students' ability to apply the basic concepts of experimental design will be enhanced by evaluating written descriptions of experiments, as well as by conducting investigations. **Through written scenarios, students can practice identifying and diagramming the basic concepts of experimental design.** Scenarios can also provide a basis for discussing typical design flaws and methods of improvement. In this chapter, five scenarios illustrating different design flaws are presented and discussed (see Activity 2.1, *Design Detective*).

COMPOST AND BEAN PLANTS

Scenario 1

DESCRIPTION: After studying about recycling, members of John's biology class investigated the effect of various recycled products on plant growth. John's lab group compared the effect of different aged grass compost on bean plants. Because decomposition is necessary for release of nutrients, the group hypothesized that older grass compost would produce taller bean plants. Three flats of bean plants (25 plants/flat) were grown for 5 days. The plants were then fertilized as follows: (a) Flat A: 450 g of 3-month-old compost, (b) Flat B: 450 g of 6-month-old compost, and (c) Flat C: 0 g compost. The plants received the same amount of sunlight and water each day. At the end of 30 days the group recorded the height of the plants (cm).

© 1998 PhotoDisc, Inc.

DISCUSSION: Use the scenario to review the basic concepts of experimental design introduced in Chapter 1. Ask the students to define the concepts and to cite examples from the scenario:

- **Hypothesis:** a prediction about the relationship between the variables that can be tested, for example, *If older compost is applied, then plant growth will be increased;*
- **Independent Variable (IV):** the variable that is purposefully changed by the experimenter, such as, *age of compost;*
- **Dependent Variable (DV):** the variable that responds, for example, *height of the bean plants;*
- **Constants (C):** all factors that remain the same and have a fixed value, for example, *amount of light and water;*

© 1998 PhotoDisc, Inc.

- **Control:** the standard for comparing experimental effects, such as, *plants in Flat C that received 0 g fertilizer;*
- **Repeated Trials:** the number of experimental repetitions, objects, or organisms tested at each level of the independent variable, such as, *the 25 plants in each flat represented 25 trials;*
- **Experimental Design Diagram:** a diagram that summarizes the independent variable, dependent variables, constants, control, number of repeated trials, experimental title, and hypothesis (see Figure 2.1 page 14).

After students have successfully defined and identified these concepts, ask them to suggest ways to improve the experiment. Initially, students may require assistance in detecting typical design flaws:

- no control group
- poorly defined constants
- insufficient number of trials
- limited number and type of dependent variables
- insufficient time span and frequency of measurements
- inappropriate use of animals

ACTIVITY 2.1 • Design Detective

SCENARIO 1 • Compost and Bean Plants

After studying about recycling, members of John's biology class investigated the effect of various recycled products on plant growth. John's lab group compared the effect of different aged grass compost on bean plants. Because decomposition is necessary for release of nutrients, the group hypothesized that older grass compost would produce taller bean plants. Three flats of bean plants (25 plants/flat) were grown for 5 days. The plants were then fertilized as follows: (a) Flat A: 450 g of 3-month-old compost, (b) Flat B: 450 g of 6-month-old compost, and (c) Flat C: 0 g compost. The plants received the same amount of sunlight and water each day. At the end of 30 days the group recorded the height of the plants (cm).

SCENARIO 2 • Metals and Rusting Iron

In chemistry class, Allen determined the effectiveness of various metals in releasing hydrogen gas from hydrochloric acid. Several weeks later, Allen read that a utilities company was burying lead next to iron pipes to prevent rusting. Allen hypothesized that less rusting would occur with the more active metals. He placed the following into 4 separate beakers of water: (a) 1 iron nail, (b) 1 iron nail wrapped with an aluminum strip, (c) 1 iron nail wrapped with a magnesium strip, (d) 1 iron nail wrapped with a lead strip. He used the same amount of water, equal amounts (mass) of the metals, and the same type of iron nails. At the end of 5 days, he rated the amount of rusting as small, moderate, or large. He also recorded the color of the water.

SCENARIO 3 • Perfumes and Bees' Behavior

JoAnna read that certain perfume esters would agitate bees. Because perfume formulas are secret, she decided whether to determine whether the unknown Ester X was present in four different perfumes by observing the bees' behavior. She placed a saucer containing 10 ml of the first perfume 3 m from the hive. She recorded the time required for the bees to emerge and made observations on their behavior. After a 30-minute recovery period, she tested the second, third, and fourth perfumes. All experiments were conducted on the same day when the weather conditions were similar; that is, air temperature, and wind.

SCENARIO 4 • Fossils and Cliff Depth

Susan observed that different kinds and amounts of fossils were present in a cliff behind her house. She wondered if changes in fossil content occurred from the top to the bottom of the bank. She marked the bank at five positions: 5, 10, 15, 20, and 25 m from the surface. She removed 1 bucket of soil from each of the positions and determined the kind and number of fossils in each sample.

SCENARIO 5 • *Aloe vera* and Planaria

Jackie read that *Aloe vera* promoted healing of burned tissue. She decided to investigate the effect of varying amounts of *Aloe vera* on the regeneration of planaria. She bisected the planaria to obtain 10 parts (5 heads and 5 tails) for each experimental group. She applied concentrations of 0%, 10%, 20%, and 30% *Aloe vera* to the groups. Fifteen ml of *Aloe vera* solutions were applied. All planaria were maintained in a growth chamber with identical food, temperature, and humidity. On Day 15, Jackie observed the regeneration of the planaria parts and categorized development as full, partial, or none.

© 2000 by Kendall/Hunt Publishing Company, Cothron, Giese, & Rezba, *Students and Research*.

Title: The Effect of Different Aged Compost on Bean Plant Growth
Hypothesis: If older compost is applied, then plant growth will be increased.

IV: Age of Compost		
3-month-old compost	6-month-old compost	No compost *(Control)*
25 Plants	25 Plants	25 Plants

DV: Height of plants (cm)
C: Amount of light
        Amount of water
        Amount of compost

Figure 2.2 Experimental Design for Plants and Compost.

In Scenario 1, students will frequently infer that the plant species and type of soil are constants. Use these inferences to illustrate the importance of explicitly stating all constants in an experiment. Ask students to describe how compost affects plant growth. Potential responses include color of leaves, number of flowers or fruits, size of leaves, and sturdiness of stems. Use these responses to distinguish between **quantitative measures,** which are based on equal interval scales, and **qualitative measures,** which are based on verbal descriptions or non-equal interval scales. Because many student researchers make insufficient measurements over a short interval, sensitize students to this pitfall. Graph hypothetical data on two groups of plants with the same average height at 24 days but different average heights at 12 days. Ask students to interpret the data and to make recommendations for fertilizer usage. Inquire about how many measurements would be necessary, over time, for them to be comfortable with the recommendation.

METALS AND RUSTING IRON

Scenario 2

© 1998 PhotoDisc, Inc.

DESCRIPTION: In chemistry class, Allen determined the effectiveness of various metals on releasing hydrogen gas from hydrochloric acid. Several weeks later, Allen read that a utilities company was burying lead next to iron pipes to prevent rusting. Allen hypothesized that less rusting would occur with the more active metals. He placed the following into four separate beakers of water: (a) 1 iron nail, (b) 1 iron nail wrapped with an aluminum strip, (c) 1 iron nail wrapped with a magnesium strip, (d) 1 iron nail wrapped with a lead strip. He used the same amount of water, equal amounts (mass) of the metals, and the same type of iron nails. At the end of 5 days, he rated the amount of rusting as small, moderate, or large. He also recorded the color of the water.

Graph: Height of plants (cm) vs. Time (days). Set A (— · —), Set B (- - - -). Y-axis: 0, 5, 10, 15, 20, 25. X-axis: 0, 2, 8, 12, 16, 20, 24.

Title: The Effectiveness of Various Metals on Preventing the Rusting of Iron

Hypothesis: If the chemical activity of the metallic wrapper is increased, then less rusting of iron will occur.

IV: Type of metallic wrapping strip			
Iron nail with no metal *(Control)*	Iron nail with magnesium	Iron nail with aluminum	Iron nail with lead
1 Trial	1 Trial	1 Trial	1 Trial

DV: Amount of rusting
Color of water

C: Amount of water
Mass of metallic wrapper
Type of iron nail

Figure 2.3 Experimental Diagram for Metals and Rusting Iron.

DISCUSSION: An excellent design feature is the presence of a control group, the iron nail without a metallic strip. The number of trials, 1 nail, is insufficient and should be increased to five or more. The number of trials required for sufficiency is determined by the variability in the experimental organisms or materials. With physical phenomena, fewer trials are necessary than for living organisms. The dependent variable, amount of rusting, should be quantified. Possibilities include measuring the residue obtained by scraping the nails or measuring the mass of the nails before and after the experiment. Both the colors of the water and the residue (precipitate) should be recorded. Color observations are critical clues to chemical reactions. In this case, they indicate the kind of iron products formed. Controlling the mass of the metallic wrapper may have introduced an undetected variable, the surface area of the iron nail exposed to the water and oxygen, into the experiment. Because chemical reactions occur at the interfaces of substances, keeping surface area constant would be a critical way to improve the experimental design that is summarized in Figure 2.3.

PERFUMES AND BEES BEHAVIOR

Scenario 3

DESCRIPTION: JoAnna read that certain perfume esters would agitate bees. Because perfume formulas are secret, she decided to determine whether the unknown Ester X was present in four different perfumes by observing the bees' behavior. She placed a saucer containing 10 ml of the first perfume 3 m from the hive. She recorded the time required for the bees to emerge and made observations on their behavior. After a 30-minute recovery period, she tested the second, third, and fourth perfumes. All experiments were conducted on the same day when the weather conditions were similar, that is, air temperature, and wind. *Note:* JoAnna's parents were experienced beekeepers and signed statements that they would monitor JoAnna's work for her safety.

DISCUSSION: Significant ways to improve JoAnna's experiment are to increase the number of trials and to add a control group such as a nonfragrant liquid (water). Caution students using animal subjects to allow sufficient recovery time between trials. Because one perfume might influence reaction to another perfume, presentation order should be randomized.

With four bee hives, a sample presentation order might be

Trial	Perfume
Trial 1	1 ..... 2 ..... 3 4
Trial 2	2 ..... 3 ..... 4 1
Trial 3	3 ..... 4 ..... 1 2
Trial 4	4 ..... 3 ..... 2 1

Students should be able to justify the constants in the experiment. "Why should the temperature be kept the same?" "Why are wind conditions important?" Students should also be able to describe the animal's normal behavior. For example, baseline data could be collected by observing the bees' frequency of emergence and behavior for several days prior to exposure to the liquids.

Students may find it strange that the 25 plants in Scenario 1 constitute repeated trials; whereas, the hundreds of bees in Scenario 3 constitute a single trial. The reason is that each plant is measured independently, giving 25 individual points of data for compost of each age; whereas, each hive is a single unit that yields only one piece of data, the time required for the bees to emerge. In Scenario 1, there are 25 points of data to average.

In Scenario 3, one cannot compute an average because only one trial was conducted (see Figure 2.4).

FOSSILS AND CLIFF DEPTH

Scenario 4

DESCRIPTION: Susan observed that different kinds and amounts of fossils were present in a cliff behind her house. She wondered if changes in fossil content occurred from the top to the bottom of the bank. She marked the bank at five positions: 5, 10, 15, 20, and 25 m from the surface. She removed 1 bucket of soil from each of the positions and determined the kind and number of fossils in each sample.

DISCUSSION: The number of trials, 1 bucket, is insufficient. Five or more samples could be drawn randomly from across the bank at the same depth. Recording both the numbers and kinds of fossils as dependent variables are an excellent feature that broadens the options for data analysis. Susan's experiment is a good example of how a nonexperimental project could be changed into an experimental project. By treating the depth from the surface as an independent variable, an **ex-post-facto experiment** or experiment after the fact was created. No control group exists in the experiment unless arbitrarily defined by the researcher. For example, if the literature re-

---

Title: The Effect of Various Perfumes on the Behavior of Bees
Hypothesis: If the perfume contains Ester X, then the bees will display agitated behavior.

IV: Type of perfume			
Perfume 1	Perfume 2	Perfume 3	Perfume 4
1 Trial	1 Trial	1 Trial	1 Trial

DV: Time to emerge
   Behavior of bees

C: Amount of perfume (10 ml)
   Distance from hive (3 m)
   Weather Conditions (air temperature, wind)

Figure 2.4 Experimental Design for Bees and Perfume.

Title: The Effect of Bank Position on Fossils

Hypothesis: As you move from the top to the bottom of the cliff, then fossil content will decrease.

IV: Depth of sample from surface				
5 m	10 m	15 m	20 m	25 m
1 Trial	1 Trial	1 Trial	1 Trial	1 Trial

DV: Kinds of fossils
       Number of fossils

C: Same amount of soil (1 bucket)

Figure 2.5 Experimental Design for Fossils and Cliff Depth.

view indicated that different types of fossils might occur above and below the 15 m level, it could be defined as the control. These designs are especially appropriate in earth science in which students are examining relationships and patterns in events that occurred long ago. This is also true for psychology where human characteristics, such as gender, intelligence, and age, are used to divide data into subgroups (see Figure 2.5).

ALOE VERA AND PLANARIA

Scenario 5

DESCRIPTION: Jackie read that *Aloe vera* promoted healing of burned tissue. She decided to investigate the effect of varying amounts of *Aloe vera* on the regeneration of planaria. She bisected the planaria to obtain 10 parts (5 heads and 5 tails) for each experimental group. She applied concentrations of 0%, 10%, 20%, and 30% *Aloe vera* to the groups. Fifteen ml of *Aloe vera* solutions were applied. All planaria were maintained in a growth chamber with identical food, temperature, and humidity. On Day 15, Jackie observed the regeneration of the planaria parts and categorized development as full, partial, or none.

DISCUSSION: The independent variable for the experiment is the concentration of *Aloe vera*.

The dependent variable is the regeneration of the planaria. A control group that received no *Aloe vera* was included; the amount of solution and growth conditions were kept the same. With living organisms, a preliminary experiment and library research should be conducted to determine the lethal dosage that should not be exceeded. Follow guidelines for humane treatment of organisms published by professional organizations. *Experimentation on vertebrates is discouraged and should be prohibited except when supervised by a mentor* (see Figure 2.6).

GENERATING EXPERIMENTAL SCENARIOS

Initially, teacher-generated scenarios are ideal because they can be constructed to focus on typical student errors. Begin by identifying the major components of the experiment and the specific design flaws.

- **Independent Variable**   Brand of car wax (4 brands)
- **Dependent Variable**   Size of water drops
- **Constants**   Same car hood
  Same amount of wax
- **Control**   None
- **Repeated Trials**   Five trials
- **Hypothesis**   Not Stated

Title: The Effect of Various Concentrations of *Aloe vera* on the Regeneration of Planaria
Hypothesis: Higher concentrations of *Aloe vera* will increase the regeneration of planaria.

IV: Concentration of *Aloe vera*			
0% *(Control)*	10%	20%	30%
10 Trials	10 Trials	10 Trials	10 Trials

DV: Regeneration of parts (heads and tails)
C: Amount of solution (15 ml)
         Growth conditions (temperature, humidity, food)

Figure 2.6 Experimental Design for *Aloe vera* and Planaria.

Include only one or two design flaws in each scenario. Remember, you want to reinforce basic principles of experimental design and sensitize students to critical ways to improve experiments. Draw an experimental design diagram to illustrate the experiment. Then, write a description of the experiments similar to the scenarios.

DESCRIPTION: Jack wanted to test which brand of car wax was most effective. He tested four brands of wax. He cleaned the hood of his car and removed the old wax. He measured four equal sections on the hood of the car. Each of the waxes was used to cover a section. An equal amount of wax, the same type of rag, and equal buffing were used. Five drops of water were placed on each square and the diameter of each drop was measured (cm).

Do not worry if students detect additional design flaws. Congratulate them for their perceptiveness. For example, students may realize that the size of the water dropper was not explicitly stated as a constant and that hood sections might have different slopes.

Investigations designed by prior students, especially if they received recognition in competitive events, make good examples. Students enjoy critiquing each others' investigations and are motivated by descriptions of what their peers have

accomplished. Experiments accompanying the textbook or curriculum can also serve as an excellent basis for class discussions. Because many textbook activities sacrifice control groups and repeated trials to save time, it is not surprising that these are the most common flaws seen in students' experimental designs. Focus students' attention on these critical components prior to conducting a textbook activity and involve the class in modifying the activity to improve the design.

MODIFYING YOUR LAB ACTIVITIES

Although students often can follow the procedural steps of a laboratory activity successfully, they frequently do not understand the purpose of the activity and its outcomes. Laboratory activities that involve manipulating one variable and observing changes in another, such as Modified Example 1.1, *A Working Heart*, in the previous chapter, are good opportunities to focus students' attention on the purpose of such activities. When getting students to read a lab activity as homework in preparation for conducting it the next day is not as successful as you wish, you might try asking them instead to complete an experimental design diagram for the lab activity. The

Title: The Effect of Water Temperature on the Heartbeat Rate of Daphnia
Hypothesis: If the temperature of the water is increased, then the Daphnia heartbeat rate will increase.

IV: Water temperature				
0℃	10℃	20℃	30℃	40℃
1 Trial	1 Trial	1 Trial	1 Trial	1 Trial

DV: Daphnia heartbeat rate (heartbeats/min)
C: Amount of water
      Waiting time
      Type of cup
      Test tube
      Slide

Figure 2.7 Experimental Design Diagram for Modified Example 1.1, *A Working Heart.*

novelty alone may be just the extra motivation needed. Figure 2.7 is an example of an experimental design diagram for Modified Example 1.1 *A Working Heart*. (See Chapter 1).

As a follow up to this experimental design diagram modification of the activity, ask students to suggest improvements to the activity, such as increasing the number of trials, designating 20°C as the control, and using the same number of Daphnia. These improvements could then be reflected in a *revised* experimental design diagram. Look for lab activities in your curriculum for which an experimental design diagram would help students understand the activity's purpose and procedures. Ask students to prepare for a lab by completing an experimental design diagram for an upcoming lab activity.

EVALUATING AND REINFORCING SKILLS

Scenarios are an excellent way to evaluate students' understanding on paper-pencil tests. Simply provide students with a brief description of an experiment and ask them to construct an experimental design diagram, identify the control (if present), and list several ways to improve the experiment. Sample practice and test items are provided in Chapters 2 and 15.

Checklists or rating sheets are also useful in helping students evaluate an experiment. After drawing an experimental design diagram of a student experiment, scenario, or textbook experiment, students can use criteria on a rating sheet to determine if they have included the following components:

© 1998 PhotoDisc, Inc.

- Title
- Hypothesis
- Independent variable
- Levels of independent variable
- Control
- Repeated trials
- Dependent variable
- Measures of dependent variable
- Constants

This checklist will also prove helpful in identifying major flaws in the experiment, such as lack of either a control group or repeated trials. Criteria may also be added to encourage students to make qualitative judgments about the merits of an experiment. For example, they can assess the appropriateness and complexity of the experiment. Is the experiment appropriate? That is, does it address a question whose answer is unknown to the student? Determining the effect of different colors of ground cover on plant growth is an appropriate experiment; whereas, determining the effect of light versus dark on plant growth is not. The student already knows what happens to plants in darkness—they die!

In determining appropriateness, students will also need to assess their own background knowledge. An appropriate experiment for a sixth grade student would not necessarily be appropriate for a high school junior or senior. Interestingly, most student experiments tend to be too complex rather than too simple. Students will tend to include too many independent variables, or attempt to investigate questions requiring more complex equipment than is available in their home or school. Frequently textbook experiments are too complex, including two or three independent variables in one experiment. Generally, these experiments can be improved by performing one experiment with one independent variable, then moving sequentially to investigations of the other independent variables.

Checklists also provide a vehicle for involving students in ongoing or formative evaluation of an experiment. After evaluating an experiment as an individual, the student can ask a peer or family member to evaluate both their experimental design and their assessment of the design. Through dialogue with others, both greater success in drawing an experimental design diagram and in evaluating the flaws of an experiment can be achieved. Finally, the teacher can offer feedback through a variety of checklists or rating sheets provided in Chapter 16. When students are first learning the concepts, you may need to simplify the task by using only a portion of a rating sheet (see Table 2.2). The first part of *Evaluating for Success 1: Basic Concepts of Design* in Chapter 16 is helpful in assessing students' understanding of the concepts in Chapters 1 and 2.

Related Web Sites

http://134.121.112.29/fair_95/gym/ index.html (abstracts for diagramming)
http://youth.net/nsrc/sci/ sci.001.html

Practice

For each of the scenarios below answer questions A–D.

 A. Identify the independent variable, levels of the independent variable, dependent variable, number of repeated trials, constants, and control (if present).

 B. Identify the hypothesis for the experiment. If the hypothesis is not explicitly stated, write one for the scenario.

 C. Draw an experimental design diagram, which includes an appropriate title and hypothesis.

 D. State at least two ways to improve the experiment described in the scenario.

1. Ten seeds were planted in each of 5 pots found around the house that contained 500 g of "Pete's Potting Soil." The pots were given the following amounts of distilled water each day for 40 days: Pot 1, 50 ml; Pot 2, 100 ml; Pot 3, 150 ml; Pot 4, 200 ml; Pot 5, 250 ml. Because Pot 3 received the recommended amount of water, it was used as a control. The height of each plant was measured at the end of the experiment.

2. Gloria wanted to find out if the color of food would affect whether kindergarten children would select it for lunch. She put food coloring into 4 identical bowls of mashed potatoes. The colors were red, green, yellow and blue. Each child chose a scoop of potatoes of the color of their choice. Gloria did this experiment using 100 students. She recorded the number of students that chose each color.

3. Susie wondered if the height of a hole punched in the side of a quart-size milk carton would affect how far from the container a liquid would spurt when the carton was full of the liquid. She used 4 identical cartons and punched the same size hole in each. The hole was placed at a different height on one side of each of the containers. The height of the holes varied in increments of 5 cm, ranging from 5 cm to 20 cm from the base of the carton. She put her finger over the holes and filled the cartons to a height of 25 cm with a liquid. When each carton was filled to the proper level, she placed it in the sink and removed her finger. Susie measured how far away from the carton's base the liquid had squirted when it hit the bottom of the sink.

4. Sandy heard that plants compete for space. She decided to test this idea. She bought a mixture of flower seeds and some potting soil. Into each of 5 plastic cups she put the same amount of soil. In the first cup she planted 2 seeds, in the second cup she planted 4 seeds, in the third cup 8 seeds, and in the fourth cup she planted 16 seeds. In the last cup she planted 32 seeds. After 25 days, she determined which set of plants looked best.

5. Esther became interested in insulation while her parent's new house was being built. She decided to determine which insulation transferred the least heat. She filled each of 5 jars half-full with water. She sealed each jar with a plastic lid. Then she wrapped each jar with a different kind of insulation. She put the jars outside in the direct sunlight. Later, she measured the temperature of the water in each jar.

© 2000 by Kendall/Hunt Publishing Company, Cothron, Giese, & Rezba, *Students and Research*.

Generating Experimental Ideas

Objectives

- Use the *Four Question Strategy* and a prompt to brainstorm numerous variables, constants, and hypotheses for experiments.
- Describe a variety of prompts for use in brainstorming: general topics, lists of materials, science articles, questions, demonstrations, and textbook laboratory activities.
- Use the *Four Question Strategy* to demonstrate the variety of options available for student experiments.
- Develop prompts for students to use with the *Four Question Strategy*.

National Standards Connections

- Identify questions that can be answered through scientific investigation (NSES).
- Design and conduct scientific investigations (NSES).

The fear of the blank page frequently paralyzes the thought processes of young would-be poets when they are asked to spontaneously write a poem. This same fear grips science students when they are asked to generate an original experiment for a class assignment or for a science fair. It does not matter how well the student has memorized the scientific method or even the definitions of variables, controls, and repeated trials. The result is always the same—panic and random thoughts. "What problem should I investigate? Why can't I think of a problem? Hey! I've got it! But, who cares? What about . . .? No, that would take too long and besides I don't have. . . ." And so it goes.

Teachers of young writers have learned to have students talk through a topic; list key ideas;

make brief notes; then write, revise, and edit. These teachers know that good writing does not begin by writing the final product. Teachers of writing have students think through a topic first by exploring various possibilities. A rough draft to get the major points on paper follows; spit and polish comes later.

Similarly, science teachers should have their students explore the possible variations of a research topic before students attempt to state a problem, a hypothesis, variables, constants, and the control. Limited class time, however, prevents most teachers from individually helping 30 or more students think through potential research problems on a myriad of science topics. Helping each student to develop a good experimental design seems nearly impossible.

Students who are assigned the task of bringing a specific research problem to class often propose broad topics such as plants or electricity. They simply do not comprehend the specificity that the phrase **specific research problem** conveys to a science teacher. Many teachers may remember that as graduate students the task of determining a specific thesis problem was very difficult. Initial draft proposals from some adults are almost as general as student proposals.

What frequently follows is the temptation of providing students with sources of well-defined projects. Some students follow the project directions as a recipe. Later, when a judge asks for the reasons a certain method was used, the student replies, "That's what the book said to do." Sound familiar?

Students need a strategy to develop a topic into a well-designed experiment. They also need that strategy modeled and practiced several times before designing an original experiment on their own. Most students are unfamiliar with both the topic area and the processes involved in conducting the research. No wonder they become easily frustrated and ask, "What do you want? I don't understand how to do this."

THE FOUR QUESTION STRATEGY

Realistically, how can one teacher help 30 to 150 students design quality original research projects and remain sane? Because the typical student's idea of a specific research problem is a general topic like plants, use it as a prompt for introducing the process to the entire class. Begin by modeling a sequence of four questions for generating experimental ideas from a general topic (plants).

Q1. What materials are readily available for conducting experiments on (*plants*)?
Soils
Plants
Fertilizers
Water
Light/heat
Containers
Q2. How do (*plants*) act?
Plants grow

Q3. How can I change the set of (*plant*) materials to affect the action?
Water
Amount
Scheduling
Method of application
Source
Composition
pH

Plants
Spacing
Kind
Age
Size

Containers
Location of holes
Number of holes
Shape
Material
Size
Color

Similarly, changes in soil, fertilizers, and light could be listed. Question 3 is the time for brainstorming; the longer the lists the better. Answering this question allows students to see the numerous possibilities, even with a simple topic.

Q4. How can I measure or describe the response of (*plants*) to the change?
Count the number of leaves
Measure the length of the longest stem
Count the number of flowers
Determine the rate of growth
Count the number of leaves
Mass (weigh) the fruit produced
Measure the diameter of the stems

At this point, conduct a discussion of which method of measurement would be best. Require students to justify both the method of measurement and the frequency of measurement.

Students can learn to write a hypothesis by relating a response to Question 3 with a response to Question 4 by using the following format: **If** I change (an independent variable from Question 3), **then** the (dependent variable from Question 4) will change. Using the class responses, have groups of students design an experiment by selecting one variable that they will deliberately change or vary . . . the **independent variable,** and one specific response that they will measure . . . the **dependent variable.** Beginning students should be limited to choosing **only one** independent variable, for example, amount of water per day, and one dependent variable, for example, the length of the longest stem. All other potential independent variables generated in response to Question 3 become **constants** for their experiment. Constants are often called controlled variables, but students find the terms controlled variables and control group confusing. Therefore, the authors prefer to use the term **control** for the group that serves as the standard for comparison, the term **variables** for those factors that change in an experiment, and the term **constants** for those factors that are kept the same in an experiment.

have the students work in small groups using the four questions as their thinking strategy. To Question 1, "What materials are readily available for conducting experiments on (**motors**)?" possible responses might be: batteries, small hobby motors, wire, string, and weights. A typical response to Question 2, "How do (**motors**) act?" is that motors lift weights. Possible student responses to Question 3, "How can I change the set of (**motor**) materials to affect the action?" might be:

Batteries	Hobby Motors	Wire	String
Brand	Brand	Length	Length
Voltage	Size	Diameter	Diameter
Age	Shape	Type	Type
Size			

To question 4, "How can I measure or describe the response of (**motors**) to the change?" student answers could be:

- Speed of lift
- Maximum weight of lift
- Number of times lifted

Continue to provide other appropriate practice topics until students are comfortable with the *Four Question Strategy* and are able to designate variables and constants in their experimental designs. See Modified Example 3.1 on page 33 for suggestions on how to use the *Four Question Strategy* with your current lab activities. When they are successful on these tasks, some students may be ready to suggest their own topics and to propose possible research designs.

Possible group choices might be:

Group 1

Independent variable	Amount of fertilizer (increments of 5 g)
Dependent variable	Height of plants
Constants	Except for amount of fertilizer, all the potential variables listed as responses to question 3 become the constants for this experiment.

Group 2

Independent variable	Amount of water (Increments of 50 ml)
Dependent variable	Number of leaves
Constants	Except for amount of water, all the potential variables listed as responses to question 3 become the constants for this experiment.

Encourage students to suggest procedures for conducting repeated trials, for example, 30 identical plants receive 5 g of fertilizer, another 30 receive 10 g, and so on. Emphasize the importance of a control, such as the 30 plants receiving 0 g of fertilizer.

APPLYING THE FOUR QUESTION STRATEGY

A class should now be ready for semi-independent practice. Using the general topic of **motors,**

PROMPTS FOR BRAINSTORMING

Other students may require more practice and a prompt or two before they are able to propose a research design. Effective prompts include:

- lists of simple and available materials;
- questions to be investigated;
- news briefs or articles that lend themselves to further experimentation by students;
- science demonstrations in a book;
- textbook or laboratory activities;
- library book.

Beginning students are most successful with lists, questions, and articles related to familiar objects, events, and organisms. Materials on each list should be selected to stimulate questions on topics related to the science subject being studied. A list of materials that includes various motor oils, a graduated cylinder, a balance, and squeaky wheels can be a very effective stimulus for physical science students. The thoughts of biology students might be sparked by a card that lists beetles, grain, insecticides, and boxes. Questions involving everyday objects or pets, such as "What shapes do pets notice most?", are effective in prompting beginners to focus on a topic for investigation.

More advanced students will be appropriately challenged by news briefs or articles suggestive of areas for further experimentation. A short article on the ability of *Chamydomonas pleichloris,* a one-celled green algae, to orient itself in response to a magnetic field raises questions like, "Do all algae respond to a magnetic field?" The article's speculative statement that ". . . the sensor could be the pyrrhotite because the polluted water in which the algae lives is rich in iron" suggests that changing the concentration of iron or varying the iron compound in the water might affect the degree of orientation to a magnetic field. Science magazines, including *The Science Teacher* from which this example came, contain excellent ideas for research topics and stimulating science news briefs.

Beginning student researchers frequently bring library books of science demonstrations and activities to class and announce, "This is what I plan to do for my project." Turn the proposed demonstration into an investigation by requiring the student to identify potential ways to vary the materials and to measure the response. Do not allow students to just demonstrate the surface tension of water in a glass by counting the pennies required to heap the water. Instead, encourage them to investigate factors that might affect how

water heaps. These could include the effect of differences in containers and in various liquids, or the influence of the mass, volume, or metallic nature of the object that is used.

Activities and demonstrations in the textbook or adopted curriculum can also be used as prompts. Ask students for ways to vary a procedure or to answer questions that emerge from an activity. For example, "What might influence the rate at which purple onion cells undergo plasmolysis in salt water—strength of the solution, temperature, or age of the onion cells?" Using the *Four Question Strategy*, allow small groups of students to brainstorm possible investigations.

Students intuitively sense that the shortest distance between two points is a straight line. They also quickly realize that brainstorming is not the fastest route to generating a topic to investigate. The temptation to list the first independent and dependent variables that come to mind, a few constants, a control, a procedure for repeated trials, and to yell out, "We're done!" is irresistible. If this happens, there are two major losses. First, consideration of only a few variables may leave undetected a large number of potential constants. Errors can result from factors that are not held the same. The second loss is even worse. Too rapid a choice of a topic precludes students from becoming aware of numerous areas of investigation that are potentially far more interesting than the one produced by an initial thought. The intellectual excitement that results from the creative act of personally discovering all sorts of possibilities is totally missed. For students who are not yet motivated by the intrinsic joys of discovery, it is wise to offer extrinsic rewards of grades or points for generating long lists of possibilities. Other students might respond positively to challenges such as, "Which group can be our champion variable finder?"

MODIFYING YOUR LAB ACTIVITIES

Experience in the laboratory can provide students with ideas for in-class mini-projects or out-of-class independent projects in several ways. For some students, even a verification activity with cookbook type procedures may spark an idea. Other students might design an independent project to follow up on an unanswered question that arises as a result of a lab activity.

ACTIVITY 3.1 • The Four Question Strategy

1. What materials are readily available for conducting experiments on (plants)?
   RESPONSE   *Soils*
   Water
   Plants
   Light/heat
   Fertilizers
   Containers

2. How do (plants) act?
   RESPONSE   *Plants grow*

3. How can you change the set of (plant) materials to affect the action?

RESPONSE *Water*	*Plants*	*Containers*
Amount	*Kind*	*Location of holes*
Scheduling	*Spacing*	*Number of holes*
Method of application	*Age*	*Shape*
Source	*Size*	*Material*
Composition		*Size*
pH		*Color*

   (Similarly, change light, soil, and fertilizer)

4. How can you measure or describe the response of (plants) to the change?
   RESPONSE   *Count the number of leaves*
   Measure the height
   Determine the rate of growth
   Count the number of leaves
   Mass (weigh) the fruit produced
   Measure the diameter of the stem

Relating the *Four Question Strategy* to Your Research Project

1. Practice the *Four Question Strategy*, using one of the prompts listed below:

   - Goldfish
   - Acid rain
   - Wood stain
   - Shampoo
   - Soil
   - Television
   - Seeds
   - Light
   - Glue
   - Plastic wrap
   - Paint
   - Birds
   - Fertilizers
   - Detergents
   - Baseballs
   - Musical instruments
   - Carbonated beverage

2. Identify a general topic that interests you. Use the *Four Question Strategy* to brainstorm potential variables and constants for experimentation.

© 2000 by Kendall/Hunt Publishing Company, Cothron, Giese, & Rezba, *Students and Research.*

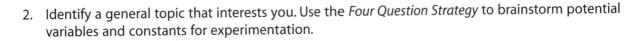

The *Four Question Strategy* and the materials cue cards described earlier can also be used. Both of these teaching techniques can be applied to many lab activities. Modified Example 3.1, *Erosion by Running Water*, for example, illustrates a typical earth science activity about erosion in which students are required to change the angle of the slope of a tray of soil and observe the resulting erosion. By using the *Four Question Strategy* with this activity, students can brainstorm numerous other variables related to erosion:

Q1. What materials are readily available for conducting experiments on *erosion*?
Water, particles (sand, silt, clay), stream tables/troughs

Q2. How does *erosion* act?
Moves materials

Q3. How can I change the set of *erosion* materials to affect the action?
Water
Amount
Velocity
Kind of spray
Temperature

Particles
Size
Kind
Density
Shape
Amount

Stream Tables
Angle
Width
Length
Material

Q4. How can I measure or describe the response of *erosion* to the change?
Weigh the amount of material moved
Determine the time required to move the material
Measure the size of the particles moved
Measure the distance moved

With this strategy, original experiments can be created by choosing an independent variable from Question 3 (keeping all others constant) and a dependent variable from Question 4. Testable hypotheses might be, "If the amount of water increased, then the amount of materials moved also increased," or "The greater the density of the rock particles, the shorter the distance moved."

Another approach to generating ideas from modified lab activities would be to completely ignore the activity directions and just use the materials list as a cue card. Direct students to design an experiment using any or all of the materials listed on the card. The variety of ideas they usually propose from the same set of materials is surprising. Consider also the number of investigations on electromagnets that might be designed using the following cue card based on the materials list from Modified Example 5.1, *Electromagnets*.

Cue Card (Erosion)	Cue Card (Electromagnets)
Stream table/trough	Batteries
Water	Different sizes an kinds of nails
Soil particles of various sizes	Insulated wire
Books	Small metal objects

Scan your own lab activities for those that might be conducive to cue cards or the *Four Question Strategy*. Look for investigations involving materials that are varied and generally available. Practice the *Four Question Strategy* on Modified Example 5.1, *Electromagnets* in Chapter 5 and see how many ways you can vary the basic materials of batteries, wire, and nails. Imagine the number of different experiments your students could do in addition to the suggested one of varying the number of coils.

EVALUATING AND REINFORCING SKILLS

Checklists or rating sheets provide another strategy for helping students evaluate a very subjective student product. For example, students can be asked to evaluate their response to Question 1, "What materials are readily available for conducting experiments on (X)?" as "excellent list," "good list," or "poor list," with the maximum score being 30 points. Initially, students will need help because there is no right or wrong answer. Involve the class in a discussion of how they would operationally define each category. "What is the minimum number of materials one should

MODIFIED EXAMPLE 3.1 • Erosion by Running Water

Purpose: To observe the effects of slope angle on the amount of erosion.

Materials: Stream table/trough
Water
Overflow bucket
Books
Sand
Clay
Pebbles

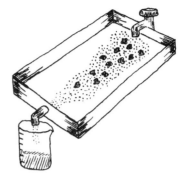

Procedure

Part I

Set up the stream table as shown. Mix equal amounts of clay, sand, and pebbles. Cover the top half of the stream table to a depth of 5 cm with the mixture. Allow the water to flow over the soil mixture for one minute. Turn off the water and observe the results.

Part II

Return the soil mixture to the top half of the stream table. Increase the slope by using books to raise the top of the stream table. Allow the water to flow as before and observe the results.

Analysis

1.   How were the different size particles moved by the water?
2.   What effect did increasing the slope of the stream table have on the amount of erosion?

Modifications for Generating Ideas for Experiments

Ask students to use the *Four Question Strategy* to brainstorm other experiments they could conduct related to erosion.

1.   What materials are readily available for conducting experiments on *erosion?*
2.   How does *erosion* act?
3.   How can I change the set of *erosion* materials to affect the action?
4.   How can I measure or describe the response of *erosion* to the change?

list to receive a rating of excellent? of good? of poor?" "How would you distribute the 30 points among the different ratings?" "What would you do if a student listed more materials than required for a rating of excellent?" "What would you do if a student listed a number of materials between the number required for poor and good?" Involvement in this discussion will help students learn to set criteria for evaluating a performance; it will also help them understand the difficulties a teacher faces in evaluating certain products, such as essay questions, lab reports, and responses to the *Four Question Strategy*. Similarly, students can determine how they would evaluate responses to Questions 2, 3, and 4.

In discussing Question 2, "How do (X) act?", help the students understand that excellent is synonymous with a scientifically correct understanding of the materials and their action. For example, a student might describe the swelling of an artificial sponge when placed in water as "the sponge grows." Does the student really mean grow, as in a living organism (sponge), or is there simply an increase in the volume or space occupied? Through questioning you can determine if the student correctly understands that the artificial sponge is expanding by absorbing water (correct/excellent answer); believes that the artificial sponge is growing just like a living sponge (incorrect/poor answer); or believes that the artificial sponge is non-living but is taking in water the same way as a living sponge (partially correct). When responses to Question 2 are not correct, refer students to appropriate sections of textbooks, encyclopedias, or other reference materials where they can learn more about the material's action and revise their answer.

If students supply the general topic also ask them to evaluate their creativity. Did they select a common topic, such as paper towels or plants, or

TABLE 3.1   Checklist for Evaluating Responses to Four Questions

Criteria/Value (100 points)	Self	Peer/Family	Teacher
Q. 1: Readily available materials (30)			
Excellent list Good list Poor list			
Q. 2: Action of materials (10)			
Excellent answer (correct) Good answer (partially correct) Poor answer (incorrect)			
Q. 3: Ways to vary materials (30)			
Excellent list Good list Poor list			
Q. 4: Ways to measure actions (20)			
Excellent list Good list Poor list			
Creativity of topic (5)			
Creativity of brainstorming (5)			

did they look for items whose response to variables is less readily known such as adhesive tapes or glues, paints and stains, or moth repellants? The second part of *"Evaluating for Success 1: Four Question Strategy"* from Chapter 16 is helpful as both a learning and an evaluative tool (see Table 3.1 *Checklist for Evaluating Responses to Four Questions*). Responses to practice exercises, extensions of textbook labs, or paper-pencil test items can also be scored using these criteria.

Helping students develop the skills needed for producing well designed experimental designs, while maintaining teacher sanity, is only one benefit of these procedures. There is also great satisfaction in helping students achieve the scientific value of longing to know and understand. Hearing students say, "I wonder what would happen if I . . .?," makes that extra effort worthwhile.

Related Web Sites

http://pointer.wphs.K12.va.us/118sci.htm (elementary)
http://sln.fi.edu/tfi/activity/act-summ.html
http;//www.ars.usda.gov/is/kids/fair/ideas.htm
http://members.aol.com/ScienzFairs/ideas.htm
http://www.pdlab.com/experiment.htm (Teacher background; written from business perspective.)
http://www.isd77.K12.mn.us/resources/cf/SciProjIntro.html (Elementary Grade Level)
http://ibms50.scri.fsu.edu/~dennisl/CMS.html (Middle School Level)
http://www.ed.gov/pubs/parents/Science
http://www.mcrel.org/resources/links/index.asp
http://www.awesomelibrary.org/science.html
http://nyelabs.kcts.org/flash_go.html
http://www.scri.fsu.edu/~dennisl/CMS/special/sf-hints.html (basic hints)
http://134.121.112.29/sciforum/guiding.html (Questions as prompts)
http://www.isd77.k12.mn.us/resources/cf/SciProjInter.html (general discussion of experimenting)
http://www.eduzone.com/Tips/science/SHOWTIP2.HTM
http://www.stemnet.nf.ca/~jbarron/scifair.html
http://www.sci.mus.mn.us/sln/tf/nav/thinkingfountain.html
http://www.exploratorium.edu/learning_studio/index.html
http://kidscience.miningco.com
http://www.waterw.com/~science/sample.html (Middle School)
http://weber.u.washington.edu/~chudler/experi.html (Human Biology)
http://www.flash.net/~spartech/ReekoScience/ReekoIndex.htm
http://ericir.syr.edu/Projects/Newton
http://www.eecs.umich.edu/mathscience/funexperiments/agesubject/age.html
http://www.eskimo.com/~billb/amasci.html
http://youth.net/nsrc/sci/sci.001.html
http://www.ksw.org.uk/physics/1_curric/curric.html

Practice

1. Use the *Four Question Strategy* to brainstorm ideas for experiments on the following topics as designated by your teacher. Save your answers to this question for use in Question 2, Chapter 4.

    A. Meal worms
    B. Insect repellent
    C. Molds
    D. Disinfectants
    E. Magnets
    F. Furniture polish
    G. Glass cleaners
    H. Bathroom cleaners
    I. Sodas
    J. Insulation

    K. Heart rate
    L. Dishwashing liquid
    M. White socks/household bleach
    N. Ice cubes
    O. Paints
    P. Kitty litter
    Q. Carpet fibers
    R. Bread dough
    S. Car wax
    T. Denture cleansers

2. Using one of the following suggested lists of materials, brainstorm ideas for an experiment.

List A	List B	List C	List D
glasses	freezer	two varieties fruit flies	peat moss
water	ice cube trays	heat source	wood and metal blocks
identical coins	salt	light source	newspaper
soaps	juices	containers	soils
hot plate	food coloring	fruit and vegetables	sand
salts	rubbing alcohol	insecticides	water
	construction paper		dilute acid
	containers		fruit
			large flat containers

3. Use one of the following hypothetical newspaper stories as a basis for brainstorming an experiment.

    Pier Point City: Fishing guide, Captain Joe Finn, reported that the fishing season is in full swing. When asked about the best way to fish, he replied, "Time is important; the hours just after sunrise and just before sunset are best. By all means," he said, "use red colored artificial lures and six inch plastic worms."

    Fiberville: Ms. A. Boss, plant manager of The Fibers Research Division, reports that braiding the cotton threads produces a stronger twine than twisting them or using them as three straight strands. She also reported that soaking the threads in starch or glue solutions before making the twine has no effect on the strength of the twine.

    Duddville: Dr. I. M. Smart issued a report today that is highly critical of U.S. education. He claims that today's students are less knowledgeable than their parents. He also claims that less than 30% of high school seniors can name four or more cabinet members. He said that most students cannot identify more than 30% of the countries on a world map. He further stated that the high school seniors are less able than their parents to successfully solve problems involving percentages or fractions.

© 2000 by Kendall/Hunt Publishing Company, Cothron, Giese, & Rezba, *Students and Research.*

CHAPTER 4

Describing Experimental Procedures

Objectives

- Write a clear and precise set of experimental procedures.
- Use a strategy to teach students the reasons for clear and precise experimental procedures.
- Use the Four Question Strategy to assist students in writing clear and precise experimental procedures.

National Standards Connections
- Communicate scientific procedures and explanations (NSES).

"**G**eorge, in your lab procedure there is a step missing between placing the acid in the beaker and beginning to add the base from the burette." "Yeah," he says upon reflection, "I put the indicator phen . . . phenol . . . phenolth . . ."

"Phenolthalein," you say.

"Yeah, that stuff, I put it in with the acid."

"Well, you didn't say so here!"

"No, but that's the point of the lab. Everybody knows that. I added it, didn't I? If I hadn't, I'd still be adding base. I know, you know, everybody knows you must add an indicator at that point, so I just skipped writing the obvious."

You know *Don't Write the Obvious George . . .!* **Getting him and many of your other students to write a detailed procedure for a scientific investigation is a major challenge.** Frequent reminders and deducting points just do not seem to work. An alternative does exist. Let the students interact and critique each other's skills of procedure writing. They will frequently do for each other what they will not do for you or for themselves. Here's how.

PEANUT BUTTER PROCEDURES

Begin by helping students perceive the importance of detailed directions by having them write procedures for something they want to do. Choose something simple so they can easily focus on the skills of procedure writing. This strategy works best when students are not yet aware that detail and thoroughness are important. The authors suggest it be done totally out of context, with no obvious connection to experimental design or procedure writing.

All students want to do the forbidden—to eat in class. The evidence for this is the candy and cracker wrappers that magically appear in desks each day despite your most practiced eagle-eye. Capitalize on this biological urge, but on your terms, not the students'. Use this opportunity to restate the rules that *there is to be no unauthorized eating in a science class or laboratory and that laboratory equipment must never be used as eating utensils.* Then cite the reason for these rules—the very real possibility of contamination of foods from laboratory surfaces or containers.

With great fanfare, cover a demonstration table with a large piece of heavy paper to serve as a contamination barrier. As a further preventative measure, insist that all utensils and food be kept on paper plates. Bring the following to each class:

- a box of crackers with inner bags unopened
- a table knife
- a jar of jelly with the lid
- a jar of peanut butter with the lid

Tell your students to put their names at the top of a sheet of paper and to individually write the directions for making a peanut butter and jelly cracker-sandwich. Allow about five minutes before collecting their papers. Ask several students to try to make a sandwich by precisely following their own directions as you read them aloud.

If a student does not describe opening the box, the inner bag, or the jar, they should sit down empty-handed and hungry. Let successive students take their turns. Do not let students shift hands, put things down, open their mouths, or do anything that their directions do not explicitly say to do. The results are hilarious and frequently resemble the party game of *Twister*. Most students will not get a cracker-sandwich. Repeat the writing exercise again. This time allow at least 10 minutes

for students to work cooperatively in small groups. It is amazing—the discussion and the level of written detail that results. And all this for a simple snack they could have at home anytime.

Compare the need for explicit detail and proper sequence for sandwich-making directions and for procedures for a science experiment. Remind students that the procedures they write must allow anyone to conduct their experiments exactly as they did them. Readers should not need to infer any details in an experimental plan, because in doing so they may vary from what the original experimenter did.

Provide your students with additional practice with one or more of the following:

- making and drinking instant ice tea
- washing a blackboard
- making a paper book cover
- tying a shoe
- tying a tie.

Note that students have experience with each of these procedures. Familiar activities make it easier for students to practice the skills of writing detailed procedures; they will in time apply these skills to the less familiar activities of science.

FROM BRAINSTORMED IDEAS TO EXPERIMENTAL DESIGN TO PROCEDURES

Although students may be able to write a procedure for a familiar task, they are not ready, either individually or in small groups, to be given a blank page, a list of materials, and the task of writing a procedure for an experiment. They must be taught how to systematically think through a new procedure. This is especially necessary when that procedure must explicitly state the variables, the constants, and the details of each step to be taken. Begin with the familiar! Use the *Four Question Strategy* to brainstorm ideas and the experimental design diagram to clarify important elements of a specific experiment. Then, prepare students to write procedures by asking them to visualize the steps necessary to conduct the experiment, identify necessary materials and equipment. Finally, have students write a set of procedures for the experiment.

BRAINSTORMING IDEAS

Begin the teaching process of procedure writing with a simple, familiar material, such as popcorn, as the basis of the *Four Question Strategy*.

Q1. What materials are readily available for conducting experiments on (*popcorn*)?
Popcorn, oil, popper
Q2. How does (*popcorn*) act?
It pops.
Q3. How can you change the set of (*popcorn*) materials to affect the action?

Popcorn
Brand
Amount
Age
Storage method

Oil
Brand
Amount
Kind

Popper
Brand
Heating time
Cooking time

Q4. How can you measure or describe the response of (*popcorn*) to the change?
Count the number of popped kernels
Measure the mass of the popped and unpopped kernels
Describe the appearance of the popcorn, e.g., color, fullness of popped kernels

CLARIFYING EXPERIMENTAL COMPONENTS

The responses to the four questions contain several parts of an experiment. Students just need to identify them. Then, they can use these components as clues to identify the other parts needed to complete the design of an experiment.

Independent Variable

Begin by having students identify an independent variable. An independent variable is the part of an experiment that the experimenter changes on purpose. When the class members responded to Question 3, they listed 10 potential ways to change the set of popcorn materials. Which one should be selected? It's up to the class, but they can only choose one. In a simple experiment there is only one independent variable. Suppose the class selects the amount of oil as their independent variable. That choice means the students will vary the amounts of oil placed with the popcorn when it is cooked.

Levels of the Independent Variable

Next, the students must decide on the levels of the independent variable. In this experiment, the levels of the independent variable are the amounts of oil tested. The students may decide to use four levels, such as 0, 10, 20, and 30 ml of oil, or only two levels, such as 0 and 30 ml of oil. Decisions about the number of levels and the amount of each level must be made. With too few levels of the independent variable, the data will not give a clear picture of the effect. Too many levels gives more data than the experimenters need. Before making these decisions, have students use cookbooks and library resources to learn as much as possible about the variables, oil and popping corn.

Control

If oil is added to each batch of popcorn in an experiment and all the batches pop reasonably well, how would the students know that it was the oil that made a difference? Maybe the corn popped well for other reasons. Maybe it might have popped better with no oil. To determine what effect the oil had, if any, a comparison group or control is needed. In this experiment, the control is the set of popcorn that received no oil (0 ml).

Dependent Variable

In what ways might varying the amount of oil affect the popcorn? In Question 4, students generated many possibilities. Any one of them could be the dependent variable. Have the class select one. If the students choose the number of kernels popped, then that option becomes the dependent variable for the experiment. The students must also decide how to operationally define a popped kernel. Will they count kernels that are partially popped or only those that are fully popped? How will each student know the difference?

Hypothesis

Remind students how to write a hypothesis. A hypothesis is a prediction of the effect that changes in the independent variable will have on the dependent variable. One possible hypothesis would be: *If I increase the amount of oil put on the popcorn, then the number of popped kernels will increase.*

Constants

If only one independent variable is needed for an experiment, why was it important to brainstorm so many responses to Question 3? Each response to Question 3 is a potential independent variable. If the students choose amount of oil as their independent variable, then *all* other potential independent variables listed under popcorn, oil, and popper must be made constants. To change a potential independent variable into a constant, assign it a specific value—amount, brand, time, and so on.

Popcorn	Value
Brand	"Pop-Rite Corn"
Amount	100 kernels
Age	1 year old
Storage	Air-Tight Container
Oil	
Brand	"Pazol"
Kind	Corn
Substitute	None
Popper	
Brand	"Cor-Pop"
Heating Time	2 min.
Cooking Time	4 min.

When the experimenter assigns a value for each constant, it is important to assign the value carefully and to choose values that are appropriate. The value of the constant should not interfere with the effects of the independent variable on the dependent variable. Suppose for the constant, methods of storage, the students assign the value "in an open jar" rather than "in an air-tight container." Improper storage may have so changed the popcorn that it will not cook properly under any conditions. Then, it will not be possible to determine how various amounts of oil affect the popcorn because of the overwhelming effect of improper storage. To make better choices for constants, have students learn as much as possible about the topic—popping corn. Consult textbooks, cookbooks, library resources, and people who live in the community.

REPEATED TRIALS

The students also need to decide how many times to gather data for each level of the independent variable. How many times will the corn be popped with each different amount of oil—1, 3, 6, or 10 times? Besides the very practical considerations of time and expense, there are two other considerations that help to define the number of repeated trials that should be conducted in an experiment. These considerations are the amount of variation in the organisms or objects being used and the consequences of coming to a wrong conclusion. Because there is apt to be more variation in popcorn kernels than there is in the measurement of oil, an experiment in which the kind of popcorn is the independent variable would need more trials than an experiment involving different amounts of oil. Because the consequences of doing an experiment to determine which brand of popcorn to buy for a movie theater chain is greater than doing it to determine which kind to buy for one's personal use, one would use a larger number of repeated trials in testing for the theater chain. The more trials used for each level of the independent variable, the more confident the class can be in the results. If the class conducts the experiment three times, there would be three trials for each level of the independent variable.

EXPERIMENTAL DESIGN DIAGRAM

A verbal description of an experiment can be very long. Use an **experimental design diagram** as a concise way to describe an entire experiment (see Chapter 2). Ask students to summarize the ex-

Title: The Effect of Various Amounts of Oil on the Number of Popped Corn Kernels

Hypothesis: If I increase the amount of oil placed on the popcorn, then the number of popped kernels will increase.

IV: Amount of oil (ml)			
0 (Control)	10	20	30
3 Trials	3 Trials	3 Trials	3 Trials

DV: Number of popped kernels

C: *Popcorn*   "Pop-Rite" corn, 100 kernels, fresh
   Oil   "Pazol", Corn
   Popper   "Cor-Pop" brand, heating time of 2 minutes, cooking time of 4 minutes.

Figure 4.1 Experimental Design Diagram.

periment they designed by drawing an experimental design diagram (see Figure 4.1).

VISUALIZING STEPS

Have the students close their eyes and visualize the steps they follow when popping corn. Allow time for each student to list the main steps on paper. Have neighboring students compare lists and make modifications. Compile a list on the chalkboard.

Step 1   Obtain popper
Step 2   Measure popcorn
Step 3   Measure oil
Step 4   Put oil in popper
Step 5   Heat popper
Step 6   Add popcorn
Step 7   Cook

In some instances, such as in Steps 1 to 3, the sequence is not important. In other instances, such as Steps 4 to 7, the sequence is critical. Modify the list to include the specific independent variable and constants identified from Question 3 and the specific response of the dependent variable that is to be measured (Question 4).

Step 1   Obtain popper   "Cor-Pop"
Step 2   Measure popcorn   "Pop-Rite Corn"
               100 kernels
Step 3   Measure oil   0, 10, 20, 30 ml
Step 4   Put oil in popper
Step 5   Heat popper   2 minutes
Step 6   Add popcorn   100 kernels
Step 7   Cook   4 minutes
Step 8   Measure   Number of
               popped kernels

By comparing the list in Question 3 with the visualized steps in popping corn, important variables, constants, and materials that were omitted may be detected. For example, an instrument for measuring the oil, a graduated cylinder, must be included. Using the lists as a guide, have each student write a step-by-step procedure. Remind them that the procedure is written for one value of the independent variable such as 10 ml of oil. A statement to repeat the steps for additional trials (Step 9, on next page) and another statement to repeat the steps for additional values of the independent variable (Step 10, on next page) should be included in a procedure.

Step 1   Obtain one "Cor-Pop" popcorn popper.

Step 2   Count out 100 kernels of fresh "Pop-Rite" corn.

Step 3   Measure 10 ml of "Pazol" oil with a graduated cylinder.

Step 4   Put the oil in the popper.

Step 5   Heat the oil for 2 min.

Step 6   Add the popcorn.

Step 7   Cook the popcorn for 4 min.

Step 8   Count the number of popped kernels.

Step 9   Repeat Steps 1–8 for three trials.

Step 10   Repeat Steps 1–9 for the other amounts of oil, e.g., 0, 10, 20, 30 ml.

After students have written draft procedures, have them work in small groups to analyze each others procedures for missing directions. For example, students might note that as a safety precaution the popcorn (Step 7) should be poured on a surface and allowed to cool before counting the number of popped kernels (Step 8).

If your students enter their experiments in a competition, check the competition's rules about procedure writing. Some competitions allow a procedure to be written as lists of steps and materials. Other competitions require the procedure to be written in paragraph form. If the competition requires a paragraph, it is easier for most students to list the materials and steps of a procedure first and then change them into a paragraph. For example, the procedure for the experiment on the effect of various amounts of oil on the number of popped kernels could be written as a paragraph:

> Ten ml of "Pazol" oil were placed in a "Cor-Pop" popcorn popper and heated for 2 minutes. One hundred kernels of "Pop-Rite" popcorn were added to the popper. The popcorn was cooked for 4 minutes. After pouring the corn on a surface and allowing it to cool, the number of fully popped kernels was counted. After allowing the popper to cool and be cleaned, the procedure was repeated for a total of 3 trials for each amount of oil, 10 ml, 20 ml, 30 ml, and 0 ml.

MODIFYING YOUR LAB ACTIVITIES

Writing a series of steps or a paragraph of procedures is a necessary but challenging aspect of designing and later preparing a report of a science experiment. Students will be more committed to the task if they are personally interested in the experiment for which procedures must be written. Use this motivation by applying the *Four Question Strategy* on one of your lab activities, similar in format to Modified Examples 3.1 and 5.1. When students have selected an independent variable and a dependent variable from the lists generated, have them write a procedure for *their* experiment. Visualizing the experiment, completing an experimental design diagram, preparing drafts, and working in groups are all good strategies to help them successfully write an appropriate procedure.

EVALUATING AND REINFORCING SKILLS

Because writing procedures involves several steps, students find the task difficult. Breaking the task into three major steps and evaluating the product at the end of each step promotes success. When students have completed brainstorming with the *Four Question Strategy*, remind them of the rating criteria for evaluating the strategy; encourage the to use these criteria to review and revise their brainstormed list. Similarly, when students complete the experimental design diagram, encourage them to use the rating criteria to evaluate their understanding of the basic concepts of experimental design. (See Table 16.1 *Designing and Generating Experiments* in Chapter 16.)

When students complete their written procedures, they will need to use a new set of criteria for evaluating their product. Helpful questions for focusing students' attention on important parts of the procedures follow.

- Is the list of steps complete?
- Are all materials and equipment included?
- Is the procedure written for only one level of the independent variable?
- Did you indicate the number of repetitions for repeated trials?

- Did you indicate the number of repetitions for each level of the independent variable?
- Are safety procedures described?
- Are the words spelled correctly? Is the grammar correct?
- Are the sentences correct? Did you begin with a capital letter and end with a period? Are the sentences complete—subject and verb included?

As students become more proficient, it will not be necessary to evaluate after each step in the process. You can use a shortened form, as illustrated below and in the rating sheet, *Evaluating for Success 2: Describing Experimental Procedures.* (See Table 4.1 *Checklist for Evaluating Procedures* and Chapter 16.)

When students have demonstrated proficiency in writing procedures for experiments with familiar topics, such as the effectiveness of various detergents in removing chocolate stains, have them write a preliminary procedure for a science experiment. This written procedure should be viewed as a first draft. Have students form groups to analyze each others procedures and to rewrite as necessary. As a final strategy, students may test the completeness of the procedures, by reading them and acting out exactly what is stated. Suggestions for practice exercises and paper-pencil test items are provided in Chapters 4 and 15. Performance measures for assessing the quality of students' written procedures are given in Chapter 16.

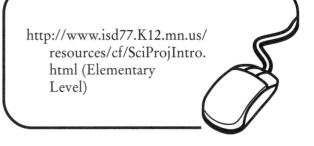

Related Web Site

http://www.isd77.K12.mn.us/ resources/cf/SciProjIntro. html (Elementary Level)

TABLE 4.1   Checklist for Evaluating Procedures

Criteria/Value	Self	Peer/Family	Teacher
All steps included (30)			
All materials/equipment included (20)			
Written for one level of independent variable (10)			
Repetitions for repeated trials (10)			
Repetitions for levels of independent variable (10)			
Written in approved format—lists or paragraph (10)			
Spelling/Grammar (5)			
Sentence/Paragraph structure (5)			

Practice

1. Write a detailed and precise procedure that includes both the sequence of steps to be taken and the materials needed for each of the following activities that is assigned to you by your teacher.

    A. You want to find out how fast the temperature of 50 ml of luke warm water would rise if it were heated by a candle.
    B. Radio station WPIG is giving away $1,000 to the first caller to get through after they hear The Pig Squeal.
    C. Your dog has fleas and you need to evaluate the effectiveness of several different brands of flea collars.
    D. You need to make a twin bed with fresh linen.
    E. You need to wax a brand new Super Hawk auto.

2. Using your answers to the practice problems in Chapter 3, draw an experimental design diagram for *one experiment* you could conduct and *write* a procedure for the experiment you developed.

© 2000 by Kendall/Hunt Publishing Company, Cothron, Giese, & Rezba, *Students and Research*.

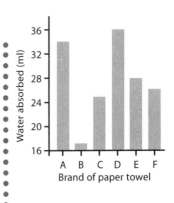

Water absorbed (ml) / Brand of paper towel

CHAPTER 5

Constructing Tables and Graphs

Objectives

- Construct an appropriate data table for organizing data.
- Construct a graph from a brief description of an investigation and a set of data.
- Draw a best-fit line for experimental data.
- Describe the relationship between variables depicted on a graph.
- Identify data for which a bar graph or a line graph is most appropriate.
- Use a structured strategy to teach students to prepare a simple data table.
- Use a structured strategy to teach students to construct a graph and to summarize trends on the graph.

National Standards Connections

- Use appropriate tools and techniques to gather, analyze, and interpret data (NSES).
- Use technology and mathematics to improve investigations and communications (NSES).
- Construct and draw inferences from charts, tables, and graphs that summarize data from real-world situations (NCTM).

"Jack, where are the data you have been collecting all period?"

"Here."

And you are shown 53 unlabeled numbers semiclustered in groups of three or four, on both sides of the page.

"How do you know which numbers are for your independent variable and which are for your dependent variable?"

"Well," says Jack, "the first trial is up here in the upper left, then the next trial is to the right, or is it this set below the first?"

"Oops," says he as he drops the paper and slowly picks it up. "Now let's see, is this the front of the sheet? Oh, never mind, I'll sort it out later. Don't worry!"

Organization and communication skills are definitely not Jack's strong suits. Perhaps the problem is Jack's, or perhaps the problem is the limited emphasis given in science classes to communication skills.

Researchers communicate information through speaking and writing, as well as through specialized vehicles including data tables and graphs. Unfortunately, many science textbooks do not emphasize these skills. Students enter numbers into pre-made data tables, plot points on preconstructed grids, and complete the fill-in-the-blank questions at the end of the lab activities. Such experiences teach students how to carefully record numbers or choose the correct term, not how to communicate. Students learn communication skills by designing data tables, constructing graphs, and formulating sentences that communicate their findings. Multiple experiences are essential for development of these higher-level thinking skills. Specific experiments and strategies for teaching data tabulation and construction of line and bar graphs are described in this chapter.

CONSTRUCTING DATA TABLES

Commercials extolling the absorbing abilities of various paper towels intrigue students. Ask students to analyze commercials to determine whether criteria for good experimental design are met, including variables, constants, and repeated trials. Challenge students to design an investigation to determine the effect of submersion time on liquid absorption or use the procedure outlined in Investigation 5.1, *Time and Absorption.* Collect data quickly by assigning different submersion times to different students, such as 10, 15, 20, 25, 30, 35, or 40 sec. Reinforce the importance of repeated trials by conducting a minimum of three trials for each submersion time. Ask the students to record their data in any way that will communicate what happened. When they have finished, ask them to evaluate their systems for recording data.

- Does the system communicate the relationship between the independent and dependent variables?
- Does the system communicate the order in which the independent variable was changed?
- Does the system's title communicate the purpose of the experiment?

Through a discussion of the students' data, introduce them to the skills of tabulating data. Although there are no universal rules for constructing data tables, generally accepted guidelines and conventions exist (see Table 5.1 *The Effect of Submersion Time on the Height a Liquid Rose in a Paper Towel Strip*). For example, the independent variable is almost always recorded in the left column and the dependent variable in the right. When repeated trials are conducted, they are recorded in subdivisions of the dependent variable column. If derived quantities (such as the average height risen per second of immersion) are calculated, they are recorded in an additional column to the right. When recording data in a table, the values of the independent variable are ordered. The data may be arranged from smallest to largest or from largest to smallest. Although no rule exists, most data are ordered from smallest to largest. The title of the data table should clearly communicate the purpose of the experiment through specific references to the variables under investigation, for example, The Effect of Submersion Time on the Height a Colored Liquid Rose in a Paper Towel. Construct a data table that illustrates these principles. Enter class data and compute the derived quantity, the average height the colored liquid rose.

EVALUATING AND REINFORCING SKILLS WITH DATA TABLES

Initially, students will need practice with both pre-collected sets of data and data from laboratory experiments. Several practice exercises and paper-pencil test items are provided in Chapters 5 and 15. Excellent practice problems are also found in the textbooks *Learning and Assessing Science Process Skills* and *Introductory Science Skills* that are referenced at the end of Chapter 5.

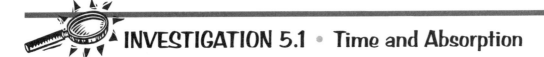

INVESTIGATION 5.1 • Time and Absorption

Materials

- Paper towel
- Food coloring
- Plastic cup or beaker
- Water
- Clock with second hand
- Scissors
- Metric ruler
- Pencil

Safety

- Handle sharp objects safely.
- Wash hands.
- Wear goggles.

Procedure

1. Cut a paper towel into strips, 2 cm x 22 cm.
2. Fill a container (cup or beaker) with water. Add several drops of food coloring.
3. Place the paper towel strip 1 cm into the colored water for the time interval designated by your teacher, for example, 10, 15, 20, 25, 30, 35, or 40 seconds.
4. At the end of each time interval, *quickly* mark the water levels with a pencil. Then, measure the height in the liquid rose in mm and record the data.
5. Repeat Steps 1 through 4 for a total of 3 trials.
6. Calculate the average height the liquid rose (mm).

Class Data Table

Construct a class data table using the following guidelines.

- Make a table containing vertical columns for the independent variable, dependent variable, and derived quantity (average height).
- Subdivide the column for the dependent variable to reflect the number of trials.
- Order the values of the independent variable—preferably from the smallest to the largest.
- Record the values of the dependent variable that correspond to each value of the independent variable.
- Calculate the derived quantities and enter the values into the table.

Column for independent variable	Column for dependent variable				Column for derived quantity
Time paper towel submerged (sec)	Height liquid rose in towel (mm) Trials				Average height (mm)
	1	2	3	etc.	

(continued on the following page)

INVESTIGATION 5.1 *(continued)*

Constructing a Line Graph

Follow the sequence described by your teacher to construct a
line graph for the class data.

- Draw and label the X and Y axes of the graph.
- Write data pairs for the values of the independent and the
  dependent variable; use the derived quantity (average
  height) for the dependent variable.
- Determine an appropriate scale for the X axis and the Y axis.
- Plot the data pairs on the graph.
- Summarize the data trends with a line-of-best-fit and
  descriptive sentences.

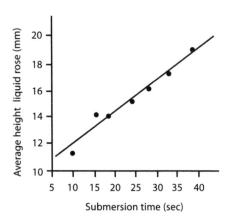

USING TECHNOLOGY ·

1. In the **STAT** mode of your calculator, enter the values for the time the paper towel was
   submerged in List 1 and values for the mean height the liquid rose in List 2. (See Appendix A, *Using Technology*, for additional help in using the graphing calculator).
2. In setting up your graph, select scatter plot as your graph type and List 1 for your
   x values and List 2 for your y values. Graph the data.
3. Examine the trend of the data and calculate a line of best fit by performing the appropriate regression (equation) analysis. For example, if the general trend of the data
   appears to be straight, you may wish to calculate a linear regression. Copy the calculated
   values to an empty Y=, and graph the equation.
4. To predict height values for submersion times that were not measured, press **Trace** and
   then the arrow keys to move along the best fit line to see the predicted y value (height)
   for any desired x value (time submerged). Depending on the brand of your calculator
   you may have to adjust the 'Window' to predict or extrapolate beyond the minimum or
   maximum x values.

© 2000 by Kendall/Hunt Publishing Company, Cothron, Giese, & Rezba, *Students and Research*.

TABLE 5.1   The Effect of Submersion Time on the Height a Liquid Rose in a Paper Towel Strip

Column for independent variable	*Column for dependent variable*			*Column for derived quantity**
Time paper towel submerged (sec)	**Height liquid rose in towel (mm)** **Trials**			**Average height (mm)**
	1	2	3	
10	11	10	11	11
15	14	14	13	14
20	14	14	14	14
25	15	15	16	15
30	16	16	16	16
35	17	17	18	17
40	19	20	19	19

*In formal data tables, the information in *ITALICS* is not included.

Ideas for modifying typical textbook activities, such as the behavior of electromagnets, to include construction of data tables are provided in Chapter 13.

Because constructing a data table involves several steps, rating sheets are useful in focusing students' attention on the process as well as serving as a tool for evaluating the product. The first part of *Evaluating for Success 3: Constructing Tables and Graphs* from Chapter 16, focuses on data tables (see Table 5.2 *Checklist for Evaluating Data Tables*).

The criteria are very objective and easy for students to score. The rating sheet can be inserted after an appropriate portion of the lab, such as completion of the data table on time and absorption in Investigation 5.1, *Time and Absorption*. Students can evaluate the table before proceeding to graphing and writing about the data.

CONSTRUCTING LINE GRAPHS

Graphs communicate in pictorial form the data collected in an experiment. Usually, a well-constructed graph communicates experimental find-ings more readily than a data table. However, graphs are more difficult to construct and involve several subskills, including knowledge of the major parts of a graph, relating data pairs from a data table to data pairs on a graph, constructing an appropriate scale for each axis, plotting the data on a graph, and summarizing the trends through a line-of-best-fit and descriptive sentences. Subskills are described later using the class data recorded in Table 5.1, *The Effect of Submersion Time on the Height a Liquid Rose in a Paper Towel Strip*.

Draw and Label Axes

Graphs are pictorial representations in two dimensions—horizontal and vertical. By convention, scientists place the independent variable, for example, time of submersion, on the X or horizontal axis and the dependent variable, for example, height the liquid rose, on the Y or vertical axis. The unit of measurement is placed in parentheses next to or beneath the variable (see Figure 5.1).

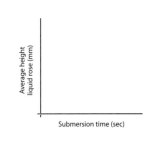

TABLE 5.2   Checklist for Evaluating Data Tables

Criteria/Value (100 points)	Self	Peer/Family	Teacher
Title (10)			
Vertical column for independent variable (10)			
Title/Unit of Independent variable included (5)			
Values of independent variable ordered (10)			
Vertical column for dependent variable (10)			
Title/Unit of dependent variable included (5)			
DV column subdivided for repeated trials (10)			
Dependent variables correctly entered (10)			
Vertical column for derived quantity (10)			
Unit of derived quantity included (10)			
Derived quantity correctly calculated (10)			

Write Data Pairs

Points on a graph are represented by a set of data. By convention, the value for the horizontal (X) axis is written first, followed by the corresponding value for the vertical (Y) axis. The two numbers are separated by a comma. Both values are placed in parentheses, for example, (10, 11). In mathematics, data pairs are called *number pairs*. If the convention for placement of the IV and DV in the data table is followed, the data pair will be in the same sequence (see Figure 5.1).

(x, y)
(10, 11)
(15, 14)
(20, 14)
(25, 15)
(30, 16)
(35, 17)
(40, 19)

Determine Scales for Axes

Determining a scale that is appropriate to cover the range of measurements for a variable is a difficult task for many students. Success has been obtained using the following procedure adapted from Rezba et al., 1995.

1. To determine an acceptable value for the axis interval, find the difference between the smallest and largest values for the variable. Obtain a reasonable number of intervals by dividing the difference by 5. Using 5 as the divisor is somewhat arbitrary but results in an appropriate number of intervals. Too many intervals crowd a graph and too few make it difficult to plot points. After dividing the difference by 5, round the resulting quotient to the nearest convenient counting number. Any number that is easily counted in multiples works well, such as multiples of 2, 5, or 10.

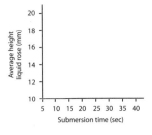

Example Scale for X axis: *submersion time (sec)*

Largest value	40 sec
Smallest value	10 sec
Difference	30 sec
Difference divided by 5	30 sec ÷ 5 = 6 sec
Quotient rounded to counting number	6 sec rounded to 5 sec

Example Scale for Y axis: *average height liquid rose (mm)*

Largest value	19 mm
Smallest value	11 mm
Difference	8 mm
Difference divided by 5	8 mm ÷ 5 = 1.6 mm
Quotient rounded to counting number	1.6 mm rounded to 2 mm

and another straight across from the 11, a point can be plotted where the two lines cross. As illustrated in Figure 5.1, use the same procedure to plot other data pairs, such as (20, 14) and (40, 19).

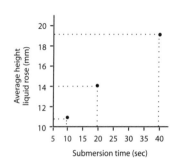

2. Develop a scale for each axis using the rounded quotient as the interval. Begin with an interval that is less than the smallest value to be graphed and end with an interval that allows the largest value to be graphed. In Table 5.1, the shortest submersion time is 10 sec so the scale for the X axis would begin with 5 sec; the scale would end with the largest value, 40 sec. Because the smallest value of the dependent variable to be graphed is 11 mm, the scale for the Y axis would begin with 10 mm; the scale would end with 20 mm to allow the largest value, 19 mm, to be graphed. Students may be accustomed to both axes on a graph having identical scales that begin with zero. Emphasize that these conditions are not necessary and may result in unused space on the graph. Instead, each scale is selected to match the data. If you use graph paper, show students how to mark the scale for each axis so that the lines on the graph paper between the intervals have a whole number value. When the scale is properly located on the graph paper, students can more accurately plot points and interpolate the values of unmeasured points (see Figure 5.1).

Plot Data Pairs

Using a data pair such as (10, 11) from Table 5.1, show students how to plot this point by locating the 10 on the X axis and the 11 on the Y axis. By sighting an imaginary line straight up from the 10

Summarize Trends

Because experimental data are subject to error, data points on a graph are not directly connected. Instead a line-of-best-fit that communicates the general data pattern is used. To construct

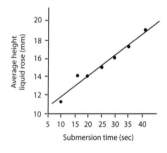

a line-of-best-fit, draw a line about which an equal number of data points fall to either side. Help students describe the trend by writing a sentence that communicates what happens to the dependent variable as the independent variable is changed, e.g., *When the amount of time a paper towel was soaked increased, the height of the colored liquid also increased* (see Figure 5.1).

Some examples of Lines-of-Best-Fit follow.

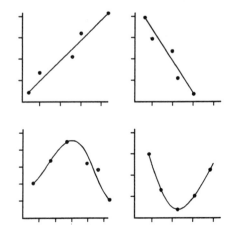

Experimental Data

Independent variable Submersion time (sec.)	Dependent variable Average height liquid rose (mm)
10	11
15	14
20	14
25	15
30	16
35	17
40	19

1. Draw and Label Axes

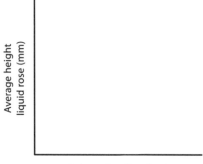

2. Write Data Pairs

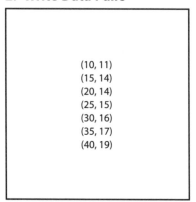

(10, 11)
(15, 14)
(20, 14)
(25, 15)
(30, 16)
(35, 17)
(40, 19)

3. Determine Scales for Axes

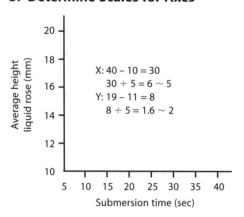

X: 40 – 10 = 30
   30 ÷ 5 = 6 ~ 5
Y: 19 – 11 = 8
   8 ÷ 5 = 1.6 ~ 2

4. Plot Data Pairs

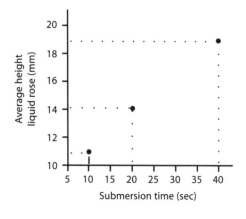

5. Summarize Trends

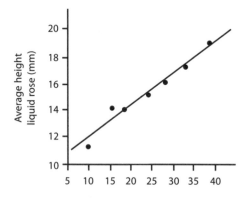

As the length of time the paper towel was submerged increased, the height the liquid rose also increased.

© 2000 by Kendall/Hunt Publishing Company, Cothron, Giese, & Rezba, *Students and Research*.

Figure 5.1 Constructing Line Graphs.

EVALUATING AND REINFORCING SKILLS WITH LINE GRAPHS

Just as with data tables, students will need practice with a variety of graphing exercises involving pre-collected data and data from laboratory experiments. For a variety of practice exercises consult Chapters 5 and 15 and the listed references. Both the line graphs resulting from these activities and from laboratory experiments can be evaluated using a simple rating sheet. The second part of *Evaluating for Success 3: Line Graphs* from Chapter 16 is very straightforward, objective, and easy for students to use (see Table 5.3 *Checklist for Evaluating Line Graphs*). When you modify laboratory activities to teach graphing, it is easy to cut-and-paste these criteria directly into the written experiment.

BAR VERSUS LINE GRAPHS

Sometimes, students are not sure whether to make a bar graph or a line graph of their data. The appropriate type of graph depends on the type of data collected. Observations and measurements of variables can be classified as either **discrete** or **continuous.** Discrete data are categorical or counted like days of the week, gender, kind of animal, brand of battery, number of children, or color. Bar graphs are appropriate for these types of variables.

Other variables are continuous and associated with measurements involving a standard scale with equal intervals. Examples include the height of plants in centimeters, the amount of fertilizer in grams, and the length of time in seconds. When the data may be any value in a continuous range of measurements, a line graph is a better way to depict the data. Line graphs allow one to interpolate or infer the value of points on a graph that were not directly measured.

There is an easy test for determining which type of graph is appropriate for a set of data. If the intervals between recorded data have meaning, a line graph is appropriate. When the intervals between the data do not have meaning, like product brands, a bar graph should be used to display the data.

CONSTRUCTING BAR GRAPHS

Because paper towel commercials strive to convince consumers of the superior absorbing ability of their product, students are easily motivated to determine which brand is superior. Ask students to design an investigation to determine the relative effectiveness of six different brands of paper

TABLE 5.3   Checklist for Evaluating Line Graphs

Criteria/Value (100 points)	Self	Peer/Family	Teacher
Title (10)			
X axis correctly labeled including units (10)			
Y axis correctly labeled including units (10)			
X axis correctly subdivided into scale (15)			
Y axis correctly subdivided into scale (15)			
Data pairs correctly plotted (15)			
Data trend summarized with line-of-best-fit (10)			
Data trend summarized with sentences (15)			

towels or use the procedure outlined in Investigation 5.2, *Brand and Absorption.* Before you begin the experiment, review the features of good data tables and involve the students in designing a class data table. Collect data quickly by assigning students different brands of paper towels.

After compiling class data, teach students to distinguish between discrete data, like brands of paper towels, and continuous data, like the amount of water absorbed. Help students understand that a line graph of the data is inappropriate because intervals between the values of the independent variable (brand of paper towel) have no meaning. Instead a bar graph is used to display the data.

Demonstrate how to construct a bar graph for the data from Investigation 5.2, *Brand and Absorption.* Use a sequence of steps similar to those outlined for line graphs (see Figure 5.2).

- Draw and label the X axis (independent variable) and Y axis (dependent variable) of the graph.
- Write data pairs for the values of the independent and dependent variables.
- Subdivide the X axis to depict the discrete values of the independent variable; that is, the six paper towel brands. Evenly distribute the values along the axis leaving a space between each value.
- Determine an appropriate scale for the Y axis that depicts the continuous values of the dependent variable, water absorbed (ml); subdivide the Y axis.
- Draw a vertical bar from the value of the independent variable (X axis) to the corresponding value of the dependent variable (Y axis). Leave spaces between each bar.
- Summarize the graph with descriptive sentences.

MODIFYING YOUR LAB ACTIVITIES

Displaying data in tables and graphs are important communication techniques in science. Because of the limitations of class time and the poor tabulating and graphing skills of many students, few authors emphasize these forms of communication. When they do, the focus is on recording data and plotting points, not on the skills of constructing appropriate tables and graphs. It is a rare lab activity that asks students to construct their own data tables and graphs. Most provide pre-constructed tables and graphs in which the independent and dependent variables are appropriately placed and labeled. In the case of graphs, the intervals also have been pre-determined and are spaced along both axes. When lab manuals contain only fill-in table and graphs, as illustrated by Modified Example 1.1, *A Working Heart* (see Chapter 1), some activities could be modified by simply eliminating these tables and graphs. The data can then be used to introduce or reinforce the desired tabulating and graphing skills.

In some texts and lab manuals, tabulating and graphing data are omitted. In those cases, lab activities can be identified that involve manipulating an independent variable and observing its effects on a dependent variable. These activities can be readily changed to include tabulating and graphing data by adding additional values (levels) of the independent variable.

An example of such a lab activity is illustrated in Modified Example 5.1, *Electromagnets* (see page 59), in which the number of wire coils around a nail is changed from 15 to 25 to determine how that affects the number of paper clips the electromagnet can pick up. This investigation could be modified by having students use additional values of 5, 10, and 20 coils and record the number of paper clips that were attracted with each new value. Repeated trials for each value of the independent variable can be conducted by individual groups, or class data can be pooled on the board. Students who were previously instructed in the conventions and rules of constructing data tables and graphs should be asked such questions as, "What are the independent and dependent variables in this lab activity?" "In a table of data, in which column are the values of the independent variables recorded?" "On which axes of a graph are the independent and dependent variables placed?" and "What is a procedure for determining the appropriate interval scales for a set of data?"

Modify your lab activities to ensure a good balance of opportunities for students to display data in tables and graphs, including those completely constructed by students. When time is limited, ask questions such as those above before allowing students to use pre-constructed tables and graphs.

INVESTIGATION 5.2 • Brand and Absorption

Materials

- Various brands of paper towels (Brands A, B, C, D, E, and F)
- Plastic container or 250 ml beaker
- Water
- Graduated cylinder (100 ml)
- Watch with second hand
- Pencil

Procedure

1. Measure 100 ml of water with the graduated cylinder. Add the water to the container (plastic container or 250 ml beaker).
2. Obtain one square of the paper towel brand designated by your teacher, such as, Brand A, B, C, D, E, or F.
3. Push the square of paper towel into the water for 30 sec. Use a pencil to push the towel under the surface.
4. Remove the paper towel. Hold the paper towel over the container until it stops dripping.
5. Use the graduated cylinder to measure the amount of water (ml) remaining in the container. Subtract the value from 100 ml to determine the amount of water (ml) absorbed by the towel.
6. Repeat Steps 1 to 5 for a total of 4 trials.
7. Calculate the average amount of liquid absorbed (ml).

Class Data Table

Construct a class data table using the following guidelines.

- Make a table containing vertical columns for the independent variable, dependent variable, and derived quantity (average water absorbed).
- Subdivide the column for the dependent variable to reflect the number of trials.
- Record the values of the independent variable (Brands A, B, C, D, E, and F).
- Record the values of the dependent variable that correspond to each value of the independent variable.
- Calculate the derived quantities and enter the values into the table.

Constructing a Bar Graph

Follow the sequence described by your teacher to construct a bar graph for the class data.

- Draw and label the X and Y axes of the graph.
- Write data pairs for the values of the independent and the dependent variables; use the derived quantities (average water absorbed) recorded in the class data table.

(continued on the following page)

INVESTIGATION 5.2 • Brand and Absorption *(continued)*

- Subdivide the X axis to depict the discrete values of the independent variable, six paper towel brands. Evenly distribute the values along the axis, leaving a space between each value.
- Determine an appropriate scale for the Y axis that depicts the continuous values of the dependent variable, water absorbed (ml); subdivide the Y axis.
- Draw a vertical bar from the value of the independent variable on the X axis to the corresponding value of the dependent variable on the Y axis. Leave spaces between each bar.
- Summarize the data trends with descriptive sentences.

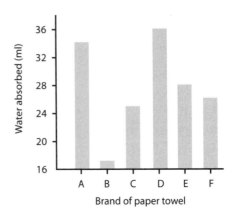

USING TECHNOLOGY •

1. In the **STAT** mode of your calculator, enter consecutive numbers (e.g., 1, 2, 3 . . .) for the brands of paper towel in List 1 and the values for amount of water absorbed in List 2. (See Appendix A, *Using Technology*, for additional help in using the graphing calculator).
2. In setting up your graph, select a histogram (bar graph) as your graph type and List 1 for your x values and List 2 as the frequency. Graph the data. Depending on the brand of your calculator you may have the option to adjust the spacing and width of the bars.

© 2000 by Kendall/Hunt Publishing Company, Cothron, Giese, & Rezba, *Students and Research*.

Experimental Data

Independent variable Brand of paper towel	Dependent variable Water absorbed (ml)
A	34
B	17
C	24
D	36
E	27
F	25

1. Draw and Label Axes

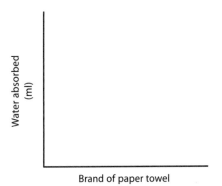

2. Write Data Pairs

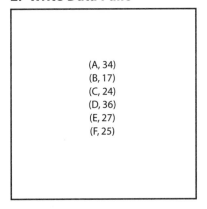

(A, 34)
(B, 17)
(C, 24)
(D, 36)
(E, 27)
(F, 25)

3. Determine Scales for Axes

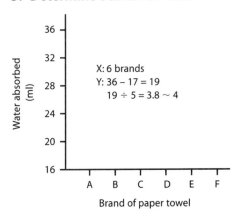

X: 6 brands
Y: 36 – 17 = 19
19 ÷ 5 = 3.8 ~ 4

4. Plot Data Pairs

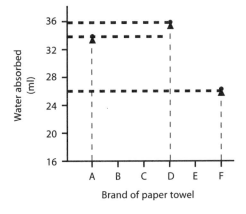

5. Summarize Trends

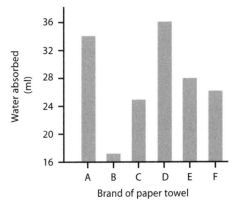

Brands A and D were the most effective water absorbers. The least effective absorber was Brand B. Brands C, E, and F absorbed intermediate amounts of water.

© 2000 by Kendall/Hunt Publishing Company, Cothron, Giese, & Rezba, *Students and Research*.

Figure 5.2 Constructing Bar Graphs.

EVALUATING AND REINFORCING SKILLS WITH BAR GRAPHS

After students have been introduced to both line and bar graphs, they will need to determine the appropriate type of graph to use with a given set of data. Practice exercises and paper-pencil test items in Chapters 5 and 15 can be used to help students.

Bar graphs resulting from these activities and from laboratory experiments can be evaluated by using the third part of *Evaluating for Success 3* from Chapter 16 that focuses on bar graphs (see Table 5.4 *Checklist for Evaluating Bar Graphs*). The criteria can be easily incorporated directly into written experiments.

After students become proficient in constructing data tables, line graphs, and bar graphs, you can use the complete rating sheet. Students may include the rating sheet in their laboratory notebook as a reference or you may post the criteria in the room.

REFERENCES

Rezba, R.J. et al. (1995). *Learning and assessing science process skills.* Dubuque, Iowa: Kendall/Hunt Publishing Company.

Gabel, D. (1993). *Introductory science skills.* Prospect Heights, Illinois: Waveland Press, Inc.

Related Web Sites

http://www.isd77.K12.mn.us/resources/cf/SciProjIntro.html (Elementary Level)

http://www.scri.fsu.edu/~dennisl/CMS/sf/sf_details.html

http://www.mste.uiuc.edu/stat/stat.html

TABLE 5.4   Checklist for Evaluating Bar Graphs

Criteria/Value (100 points)	Self	Peer/Family	Teacher
Title (10)			
X axis correctly labeled including units (10)			
Y axis correctly labeled including units (10)			
X axis correctly subdivided—discrete values (15)			
Y axis correctly subdivided into scale (20)			
Vertical bars for data pairs correctly drawn (15)			
Data trend summarized with sentences (20)			

MODIFIED EXAMPLE 5.1 • Electromagnets

Purpose: To observe the behavior of an electromagnet.

Materials: One 1.5 volt D-cell battery
Switch
Battery holder
Large nail
Insulated #20 wire, one long piece (50 cm) and one short (10 cm), stripped at
     both ends.
Steel paper clips

Procedure

1. Wrap the longer wire around the nail 15 times as pictured.
2. Attach one end of the longer wire to a switch and the other to
   a battery holder. Connect the other end of the battery holder
   to the other side of the switch with the shorter wire.
3. Close the switch and see how many paper clips you can pick up.
4. Increase the number of wraps of wire around the nail to 25 and
   see how many paper clips you can pick up now.

Analysis

1. How did you make your electromagnet stronger?
2. How do you know it was stronger?

Modifications for Constructing Tables & Graphs

Add these to the procedures:

1. Collect additional data on the behavior of electromagnets by determining the number of
   paper clips you can pick up when the number of wraps of wire is changed, such as 5, 10, 15,
   20, and 25 wraps.
2. Construct a data table and enter the data collected. (Hint: Remember repeated trials.)
3. Construct an appropriate graph to communicate the relationship between the number of
   wraps of wire and the strength of the magnet.

Practice

1. For each experiment title listed, state whether the experiment should be graphed as a bar or a line graph.

   A. The Effect of Coloration on the Number of Kittens Sold at a Pet Store.
   B. The Effect of Concentration of Sugar Water on the Number of Visits of Hummingbirds to a Feeder.
   C. The Effectiveness of Different Brands of Paper Towels on the Absorption of Water.
   D. The Effect of the Horsepower of a Tractor on the Mass of a Sled it Can Pull.

2. Construct a data table and an appropriate graph for the following sets of data.

   A. Relationship Between Distance Below Surface and Number of Fossils Collected

Distance below surface (cm)	Number of fossils collected
80	0
140	2
200	8
260	15
320	32

   B. In 1988, the USA imported food. We imported in billions of dollars the following amounts: Shellfish, $2.7; Coffee, $2.5; Beef and Veal, $1.7; Pork, $0.9; Orange juice, $0.6; Cheese, $0.4; Grapes, $0.3; Tomatoes, $0.2.

   C. Bill and Sheri decided to study possible relationships involving abdominal muscle strength and endurance. Bill chose to see whether the number of situps that male athletes who are in top shape could do in two minutes is related to their ages. Sheri chose to see whether the weight of female athletes in top shape is related to the maximum number of situps they could do in two minutes. Construct data tables and graphs for Bill's and Sheri's data.

Bill's Data				**Sheri's Data**		
Subject	Age (years)	Number of situps		Subject	Weight (in pounds)	Number of situps
Mark	14.5	95		Gail	100	71
Bob	17	100		Margo	170	59
Ron	15.5	97		Joyce	180	57
Don	18	102		Lori	110	69
Lou	14	93		Cathy	130	66
Armand	16.5	99		Dena	120	68
Norm	15	9		Tammy	150	62
Bill	17.5	101		Agnes	160	61
Doug	16	98		Linda	140	64

3. Describe the relationship between the variables graphed in Questions 2 A–C.

© 2000 by Kendall/Hunt Publishing Company, Cothron, Giese, & Rezba, *Students and Research*.

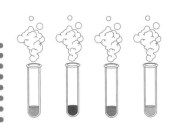

CHAPTER 6

Writing Simple Reports

Objectives

- Identify the elements of a simple report—title, introduction, experimental design, procedure, results, and conclusion.
- Write a simple report of a scientific investigation.
- Use a structured format to teach students to write a simple report.
- Enhance students' writing skills by incorporating a variety of writing opportunities into the curriculum.

National Standards Connections

- Develop descriptions, explanations, predictions, and models using evidence (NSES).
- Think critically and logically to make the relationship between evidence and explanations (NSES).
- Make inferences and convincing arguments that are based on data analysis (NCTM).

Students are frequently asked to write, but only rarely are given sufficient practice to develop the skills necessary to write well. Occasions for practicing technical or scientific writing skills are even more limited. Students can learn to write the critical elements of a formal scientific paper by learning to write simple reports consisting of six components:

1. Title
2. Introduction
3. Experimental design diagram
4. Procedure
5. Results
6. Conclusion

Because students have previously learned to write a title for an experiment, to draw an experimental design diagram, to write a clear and precise procedure, and to depict results through tables and graphs, the simple report can be formed by combining these skills with two new skills—writing an introduction and a conclusion. Furthermore, the simple report can be approached in a stepwise fashion that breaks the task into manageable components for inexperienced technical writers.

SIMPLE REPORTS

Begin with a simple experiment that allows students to readily apply their skills of experimental design and data analysis. Advertisements for household products pose numerous questions for investigation. Discuss with students, for example, advertisements for effervescent tablets that reduce stomach acid. Highlight the behavior of the tablets in water—plop, plop, fizz, fizz—and challenge them to determine whether the temperature of the water affects tablet solubility.

Ask the students to design an investigation to test the hypothesis that **if** the temperature of the water is increased, **then** effervescent tablets will dissolve faster. As an alternative, you may use the procedure outlined in Investigation 6.1, *Super Fizzers.* To save time and materials, assign a specific water temperature to each student group and create repeated trials by pooling group data into a class data table. Use ice water, room temperature water, and hot water ~ 45°C as the cold, room, and hot water temperatures needed in the investigation.

As part of the investigation, require students to write a title, draw an experimental design diagram, make a data table, and construct an appropriate graph. When students have completed the investigation, outline the major components of a simple report (see Table 6.1 *How to Write a Simple Report*). Ask them to identify familiar parts of the report. Students should recognize that only two parts—the introduction and the conclusion—are unfamiliar. Emphasize that students already know how to write four of the six components of a simple report. Through a stepwise process, they will learn to write two new elements and to combine the six components into a simple report.

Title

A good title relates the independent and dependent variables that were investigated. Initially, students may write a title using the structured format of the effect of the independent variable on the dependent variable.

Example:
The Effect of Water Temperature on the Dissolving Time of Effervescent Tablets

Introduction

In this component, the student establishes the research problem by stating the rationale, purpose, and hypothesis for the study. A proven technique for helping students compose an introduction is for them to write responses to each of the following questions. Some teachers eliminate the rationale question because many beginning students have difficulty distinguishing a rationale from a purpose.

Q1. Why did you conduct the experiment? *(Rationale)*
Some over-the-counter drugs are effervescent tablets that must be dissolved in water before ingestion. Directions do not include specific water temperatures.

Q2. What did you hope to learn? *(Purpose)*
The purpose of this experiment was to determine the effect of water temperature on the time required for effervescent tablets to dissolve.

Q3. What did you think would happen? *(Hypothesis)*
The researcher hypothesized that if the temperature of the water is increased, then effervescent tablets will dissolve faster.

Through the hypothesis, the experimenter communicates the anticipated effect of the independent variable on the dependent variable. A simple hypothesis may be written as an **if . . . then** statement, such as, "**If** soils contain a high percentage of clay (independent variable), **then** percolating rainwater will be more acidic (dependent variable)." More experienced student researchers may employ alternative sentence structures, for example, Daycare promotes aggressive behavior in toddlers as reflected by color selection. An explanation of how the student became interested in the study may also be included in the introduction.

Experimental Design

As discussed in Chapters 1 and 2, the experimental design can be communicated through a simple diagram. By convention, the diagram is based

INVESTIGATION 6.1 • Super Fizzers

Directions

Read the list of materials and the procedure for conducting the investigation. Before you begin the experiment: (1) write a title, (2) draw an experimental design diagram, and (3) construct a data table. After completing the experiment, construct an appropriate graph.

Materials

- Plastic cups
- Water at three different temperatures (ice cold, room temperature, hot)
- Brand X effervescent tablets
- Watch with second hand
- Graduated cylinder
- Goggles

Safety

- Wear goggles.
- Handle hot objects carefully.

Procedure

1. From the central supply area, obtain 75 ml of ice water.
2. Add one effervescent tablet.
3. Record the time (sec) for the tablet to completely dissolve. Discard the solution as directed by your teacher.
4. From central supply, obtain 75 ml of room temperature water. Repeat Steps 2–3.
5. From central supply, obtain 75 ml of hot water. Repeat Steps 2–3.
6. To create repeated trials, record your group's data on the class data table.
7. Compute the average time for dissolving at each temperature using the values from the class data table.
8. Construct an appropriate graph of the data.

Writing a Simple Report

Follow the sequence described by your teacher to write a simple report for this investigation.

1. Title
2. Introduction
3. Experimental Design Diagram
4. Procedure
5. Results
6. Conclusion

USING TECHNOLOGY ·

1. In the **STAT** mode of your calculator, enter consecutive numbers (e.g., 1, 2, 3) for the temperatures in List 1 and the values for mean dissolving time in List 2. (See Appendix A, *Using Technology*, for additional help in using the graphing calculator).
2. In setting up your graph, select a histogram (bar graph) as your graph type, List 1 for your x values (temperatures) and List 2 (dissolving time) as the frequency. Graph the data. Depending on the brand of your calculator you may have the option to adjust the spacing and width of the bars.

© 2000 by Kendall/Hunt Publishing Company, Cothron, Giese, & Rezba, *Students and Research*.

TABLE 6.1   How to Write a Simple Report

Part	Purpose
1. Title	Write a sentence that relates the independent and dependent variables that were investigated.
2. Introduction	Describe the rationale, purpose, and hypothesis for the investigation. Use three questions to guide your writing of the introduction. • Why did you conduct the experiment? (Rationale) • What did you hope to learn? (Purpose) • What did you think would happen? (Hypothesis)
3. Experimental Design Diagram	Format the experimental process. • Begin the diagram by drawing a rectangle. • Write the independent variable (IV) across the top of the rectangle. • Divide the rectangle into labeled columns to represent the different levels of the independent variable. • Indicate the number of trials in each column. • Write the dependent variables (DV) and constants (C) beneath the rectangle.
4. Procedure	List the steps followed to complete the investigation. Check the list carefully for accuracy, completeness, and precision.
5. Results	Complete a data table and an appropriate graph for the data using the following guidelines.
Data Table	• Make a table containing vertical columns for the independent variable, dependent variable, and derived quantity. • Subdivide the column for the dependent variable to reflect the number of trials. • Order the values of the independent variable—preferably from the smallest to the largest. • Record values of the dependent variable. • Compute the derived quantity.
Graph	• Draw and label the X and Y axes of the graph. • Write data pairs for the independent and dependent variables. • Determine an appropriate scale for the X and Y axes; subdivide the axes. • Plot the data pairs on the graph. • Summarize the data trends on the graph.
6. Conclusion	Describe the purpose, major findings, an explanation for the findings, and recommendations for further study. Use six questions to guide your writing of the conclusion. • What was the purpose of the experiment? • What were the major findings? • Was the hypothesis supported by the data? • How did your findings compare with other researchers or with information in the textbook? • What possible explanation can you offer for the findings? • What recommendations do you have for further study and for improving the experiment?

© 2000 by Kendall/Hunt Publishing Company, Cothron, Giese, & Rezba, *Students and Research*.

upon a rectangle. The independent variable (IV) is written across the top of the rectangle that is divided into labeled columns to represent the different values of the independent variable. The number of trials is indicated in the columns. The dependent variables (DV) and constants (C) are written beneath the rectangle.

Example

IV: Temperature of Water		
Cold	Room	Hot
5 Trials	5 Trials	5 Trials

DV: Time to dissolve (sec)

C: Brand of effervescent tablet
Amount of water (75 ml)
No stirring
Type of cup

Procedure

In formal papers, the experimental procedure is described in paragraph form. Inexperienced researchers find it difficult to write a clear, precise procedure in this format. Generally, they are more successful when allowed to list the materials used in the investigation and to list the steps followed in performing the investigation (see Investigation 6.1, *Super Fizzers*). Strategies for helping students write clear precise procedures are described in Chapter 3. Select the most appropriate format, paragraphs or lists, for your students.

Example:

One Brand X effervescent tablet was placed in a plastic glass containing 75 ml of cold water and the dissolving time recorded in seconds. Five trials were conducted with cold water. Similarly, dissolving times were determined in room temperature and hot water. The mean dissolving time at each temperature was calculated and an appropriate graph constructed.

Results

Data collected from an experiment are displayed in simple data tables and appropriate graphs. In constructing a data table, students make columns for the independent variable, dependent variable, and derived quantity (average dissolving time). Repeated trials are shown in subdivisions of the dependent variable column. Depending upon the type of data, continuous or discrete, students may construct a line graph or bar graph to display the data.

Graphs can be constructed by following the five stage sequence of (1) drawing and labeling the X and Y axes of the graph, (2) writing data pairs for the independent and dependent variables, (3) determining appropriate scales for the X and Y axes and subdividing the axes, (4) plotting the data pairs on the graph, and (5) summarizing the data trends. Structured strategies for teaching students to construct data tables and graphs are described in Chapter 5, *Constructing Tables and Graphs.*

Data Table							Bar Graph
	Time to dissolve (sec)					**Average time to dissolve (sec)**	
	Trials						
Temperature of water	**1**	**2**	**3**	**4**	**5**		
ice	98	104	107	96	105	102	
room	43	35	46	46	30	40	
hot	24	27	19	19	27	23	

Conclusion

Writing a conclusion requires a thinking process that simultaneously involves a student in both the analysis of data and the synthesis of relationships. As with the introduction, an organizational framework for guiding this thought process is frequently needed. Questions that have proved helpful to students and an illustrative response, are described here.

Q1. What was the purpose of the experiment?
The purpose of the experiment was to determine the effect of water temperature on the dissolving time of Brand X effervescent tablets.

Q2. What were the major findings?
At higher temperatures, Brand X tablets dissolved faster. For each rise in temperature, dissolving time was further reduced.

Q3. Was the research hypothesis supported by the data?
The data supported the research hypothesis that effervescent tablets would dissolve faster in warm water than in cold water.

Q4. How did your findings compare with other researchers or with information in the textbook?
The findings agree with the solubility rule that solids dissolve faster in warm solvents.

Q5. What possible explanations can you offer for the findings?
Because molecules move faster in warm water, they would strike the tablet more frequently and tear it apart more quickly.

Q6. What recommendations do you have for further study and for improving the experiment?
Additional experiments could be conducted to determine the dissolving rate of other brands. The experiment could be improved by insulating the cups to reduce heat exchange with the room.

Initially, students may be overwhelmed if they are presented with the six questions and required to write the entire paragraph at once. Instead, approach the task one-step-at-a-time. Divide the chalkboard into a section for questions and a section for responses. Write the first question on the board and provide students with several examples that illustrate different ways to write a purpose, such as:

- In this experiment, the effect of kind of fertilizer on the size of geranium flowers was studied.
- The researchers tested the effect of temperature on the heart rate of daphnia in this investigation.
- The purpose of this experiment was to show how amounts of $CaCl_2$ affected the temperature of H_2O.

After discussing various examples of a purpose statement, allow students several minutes to complete a response. Choose several students who composed different, but acceptable, sentences to write their sentences on the board. Emphasize that there are many ways to write the purpose as long as all of the important information is communicated. Because the introduction contained a statement of purpose, students find it easy to complete this component of the conclusion. The high success rate will help establish a positive climate for the remaining steps.

Students have more difficulty describing the major findings. Discuss several examples of sentences that summarize the major findings of experiments, such as:

- When increasing amounts of salt were dissolved in H_2O, no increase in temperature was found.
- Some runway surfaces (sand, paper, flagstone) slowed the car's speed, while others increased the speed (glass, formica) over the control surface.
- The temperature of H_2O increased as the amount of $CaCl_2$ increased.

Ask students to write their major findings. When you select sentences to be recorded on the board, include sentences that repeat the data and that summarize the data. In this way, students will feel positive about their writing, yet experience more succinct ways of stating the findings. Many students will still simply repeat the average dissolving time at each temperature, Brand X tablets dissolved in 102 sec in cold water, 40 sec in room temperature water, and 23 sec in hot water. Such sentences are typical of students in the concrete developmental stage. Be satisfied with these sentences initially but encourage student to analyze and state trends by responding orally to questions such as, "What pattern do you see?" "Did the dissolving time increase or decrease as the temperature of the water increased?" or "How much did the dissolving time change with each change in temperature?"

Next, have students write a simple sentence indicating the extent that the data supported the hypothesis. Remind students that an experiment neither proves the hypothesis nor indicates that it is correct. Revise sentences to eliminate these phrases and to include the more appropriate statement, support or did not support the hypothesis.

Comparing one's experimental findings with other research and proposing an explanation for one's research findings are very difficult for students. You may need to prompt students by referring them to specific sections of textbooks, library resources, or by asking leading questions: "Read this page in Chapter 7," "What did you learn last week about the dissolving of solids?" "Look in the encyclopedia or a chemistry textbook to learn more about the kinetic molecular theory." "In the last classroom activity, what made solids dissolve faster?"

Finally, encourage students to suggest improvements and additional investigations. Remind students that all suggestions are appropriate as long as they are consistent with the principles of experimental design or result from unanswered questions in the study. Again, you may need to ask students probing questions to elicit a response. "Did you have sufficient trials?" "Did the water temperature change during the experiment?" "Do you think all brands of tablets will dissolve at the same rate?" "What would happen if you pulverized the tablets?" "Do you think changing the temperature of the tablets instead of the water would make a difference?"

MODIFYING YOUR LAB ACTIVITIES

Most lab activities do not ask for the kind of writing that develops technical writing skills—skills that students need to adequately describe an original experiment of their own or to write reports in future workplaces. For each of the lab activities presented in these sections, such as Modified Examples 1.1, *A Working Heart*, 3.1, *Erosion*, 5.1, *Electromagnets*, and 8.1, *Pupillary Reflex*, some form of technical writing could be incorporated. For example, writing about Daphnia would be good practice for writing background information for an introduction, whereas data on pupillary reflex could form the basis for writing paragraphs of results and a conclusion. Writing a new set of procedures would be a natural extension for the experiments resulting from applying the *Four Question Strategy* to either the Erosion or Electromagnets investigations.

Additional suggestions on how lab activities can be used to improve students' writing skills in science are provided in Chapter 9, *Communicating Descriptive Statistics,* and Chapter 12, *Preparing Formal Papers.* Select a few of your labs to incorporate at least one kind of writing. Space these writing modifications over several months to help students develop skills in writing a title, hypothesis, introduction, background section, paragraph of results, and a conclusion.

EVALUATING AND REINFORCING SKILLS

If rating sheets were used to assess student products described in the first five chapters, students would be familiar with the criteria for evaluating an experimental design diagram, procedures, data tables, and line or bar graphs. Different criteria will be required for the two remaining sections of the report—the introduction and conclusion. This is a good time to introduce students to a rating sheet, such as *Evaluating for Success 4: Writing a Simple Report* from Chapter 16. After students complete each part of the report, such as the title/introduction, they can evaluate the section using the criteria listed and make revisions.

Title/Introduction (16 points)

Criteria/Value	Self	Peer/Family	Teacher
Title/Introduction (16)			
Correct title (4)			
Rationale (4)			
Purpose (4)			
Hypothesis (4)			

Then, they can proceed to the next component, the experimental design.

Experimental Design (16 points)

Criteria/Value	Self	Peer/Family	Teacher
Experimental Design (16)			
Name/Level/Units of independent variable (3)			
Control (3)			
Repeated trials (3)			
Name/Units of dependent variable (3)			
Constants (4)			

Similarly, students can evaluate the procedures, data tables, graphs, and conclusions using the appropriate section of the rating sheet.

If students are initially overwhelmed with the entire rating sheet, you can cut-and-paste the individual sections. If the criteria are too complex, you can also revise. For example, you might reduce the six-step conclusion to a three-step process and adjust the points accordingly.

• What was your experiment about?
• What did you learn in the experiment?
• How could you improve the experiment?

PRACTICING WRITING SKILLS

Writing an entire report is a complex task. Few students will have all the necessary skills for success. Most students would view the total task as insurmountable. Teachers can help by providing opportunities to write one component at a time and by using cooperative learning strategies to generate a complete report.

ONE COMPONENT AT A TIME

To encourage students to read a lab activity in advance, some teachers require students to write the steps in the procedure. When this becomes a routine assignment, it often reverts to a mindless listing of steps. Therefore, to encourage students to prepare for a lab and to improve writing skills, a variety of other writing assignments can be made.

Identifying the hypothesis to be tested and writing it in a correct form is an interesting alternative. If the lab involves manipulating variables, students can complete an experimental design diagram as described in Chapters 1 and 2. Even summarizing an experimental procedure can be made more challenging by limiting the length of the summary. To meet the limitation imposed, students will have to make decisions about essential information versus supportive detail.

Later, when students have learned the basic skills of technical writing, let them choose the kind of writing they do in preparation for a lab. Some might choose to draw an experimental design diagram, others might summarize the procedure, and still others might write the introduction or suggest improvements to the lab activity, for example, add a control, increase the sample size. Given a choice, more students may thoughtfully complete the assignment. The result will be better lab preparation and more writing practice.

Similarly, students could report results through a variety of formats—data tables, graphs, or a concluding paragraph. Recommendations for improvement or for further study could be written at other times. The key is providing a variety of thought-provoking writing opportunities. Emphasizing each report component separately makes the writing task more manageable. Varying the writing assignments from lab to lab also provides novelty that is more motivating than is repeating the same activity.

COOPERATIVE LABORATORY REPORTS

An alternative approach to teaching writing skills involves the strategies of cooperative learning. Groups of students can share the reporting task by writing different parts of the report of their jointly conducted activity. If students work in groups of four, one student can prepare the title, introduction, and experimental design diagram. Another can describe the investigative methods, the equipment, and the materials used. The third student can organize the experimental data in tables and graphs. The student who writes the final part of the report would interpret the results by drawing conclusions and stating support for the original hypothesis. Maximize each student's development by rotating the writing task among group members.

As another cooperative writing activity, groups of students could practice writing an abstract of their reported investigation. An *abstract* is a concise summary of the investigation that includes the purpose of the investigation, the hypothesis tested, a brief description of the procedure, major findings, and the conclusion. Because length restrictions are severe, usually one page or about 250 words, numerous opportunities exist in this exercise for sorting essential from nonessential information. Sometimes the choices are clear, other times they appear arbitrary or even capricious. Individuals within the group who are less proficient in making judgments about the relative importance of information will benefit from their experiences in these group decision-making sessions.

FEEDBACK AND REVISION ARE ESSENTIAL

Whenever writing is assigned as part of a laboratory activity, it should be followed with a class discussion of the critical elements of the report component. For example, when writing a procedure, some students skip critical steps, whereas others include so many unnecessary details that the essential steps become lost. Opportunities to revise should follow. By using feedback from other class members to improve their procedure, students learn to write better. Students should also see rewriting as a natural part of the writing experience. Most professional scientists rewrite papers many times before publishing them. As a final writing activity, have groups of students edit their work. At this point, grammar, punctuation, and spelling become important.

Unlike speaking, writing offers students the opportunity to cross-out mistakes and start over, to stop and think, to reformulate and revise. The instructional sequence of experimentation, analysis, discussion, writing, feedback, revision, and editing leads to the kind of attention to detail and scrutiny necessary for good writing. Well-written papers, whether competitive winners or not, are the reward for the extra effort.

Related Web Sites

http://www.dade.k12.fl.us/us1/science/prod03.htm

http://www.isd77.K12.mn.us/resources/cf/SciProjIntro.html (Elementary Level)

http://www.access.digex.net/~schapman/classes/labs.html

http://www.eduzone.com/Tips/science/SHOWTIP2.HTM (report section)

PART TWO

Advanced Principles
of Experimental Design
and Data Analysis

✓ Expand inquiry skills to include more sophisticated data analysis, experimental design, and reporting:

- use library resources
- apply descriptive statistics
- write formal papers
- design multi-variable experiments
- establish statistical significance

CHAPTER 7

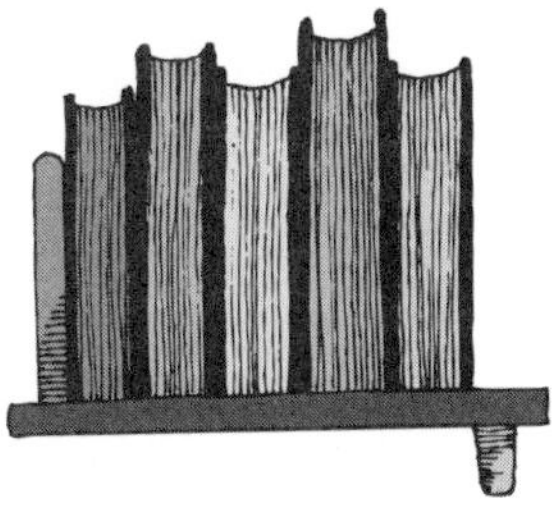

Using Library Resources

Objectives

- Identify critical information for documenting a source and use an approved style format for documentation.
- Identify critical information to record on note cards to summarize different types of information.

general sources	technical manuals and procedures
scientific research	interviews with community members

- Complete note cards on a variety of sources.
- Use library resources and questions to generate and refine ideas for an independent research project.
- Use structured procedures to teach students how to take notes from a variety of sources and to prepare appropriate note cards.
- Demonstrate how library resources can help to refine ideas for scientific research.
- Identify library and community resources appropriate for student research projects.

National Standards Connections

- Identify questions and concepts that guide scientific investigations (NSES).

To students, the most difficult part of research is selecting a topic, whereas the most frustrating part for both parents and students is accessing relevant library materials. Contributing factors include inadequate library research skills, limited library resources, and the students' nonscientific backgrounds. Add an unfocused topic, youthful impatience, deadlines, and students begin to view the library as an adversary, not an ally, in scientific research. Contrast this negative attitude with that of professional scientists, who recognize the ongoing and symbiotic relationship between library and scientific research. **Teachers can make this positive relationship a reality for students by using a five-stage model to systematically develop library skills critical to various stages of scientific research** (see Table 7.1 *Five-Stage Model for Relating Library and Scientific Research*).

TABLE 7.1 Five-Stage Model for Relating Library and Scientific Research

Stage	Use appropriate library resources	Use library research skills	Relate library & scientific research
1. Establish an interest	• Popular magazines • Newspapers	• Documenting • Scanning • Making note cards	• What **general topic** (X) interests you? • What **general action** of (X) interests you?
2. Narrow the topic	• Textbooks, K–12 • General references: books, encyclopedias, dictionaries, handbooks	• Using descriptors • Locating information in books, card catalogs, general indexes • Documenting • Scanning • Making note cards	• What **specific topic** (X) interests you? • What **materials** are readily available for you to conduct experiments on (X)? • What **specific action** of (X) interests you? • How could you **measure** or **describe** the action of (X)?
3. Clarify the variables (Optional)	• Scientific indexes • Scientific abstracts • Scientific journal articles	• Using scientific indexes • Documenting • Scanning • Making note cards	• What **variable** will **you change** to conduct experiments on (X)? • What **specific changes** will you make in (X)? • What **action** of (X) will you **investigate?** • What **specific observations** or **measurements** will you make on (X)?
4. Refine the procedures (Optional)	• Laboratory manuals, K–12 • Handbooks and manuals • Sourcebooks • Community resources	• Using community resources • Conducting interviews • Documenting • Scanning • Making note cards	• What specific **materials** or **organisms** will you use? • What specific **procedures** will you follow? • How will you **collect** and **analyze data?**
5. Interpret the unexpected	• Appropriate resources	• Using library skills	• How can you **explain unexpected** events?

© 2000 by Kendall/Hunt Publishing Company, Cothron, Giese, & Rezba, *Students and Research.*

Stage 1

ESTABLISH AN INTEREST

Generally, scientists investigate topics in their area of expertise and interest. Similarly, students will design and execute better research in an area of their interest. Use knowledge of a student's favorite school subject, hobbies, extracurricular activities, and career goals to identify potential topics (see Activity 17.1).

To determine students' scientific interests, ask them to describe a science-related newspaper article, magazine, or book they recently read. During the discussion, identify popular science magazines that are readily available in school or community libraries. If possible, have copies of these magazines available in the classroom. Include such magazines as *Smithsonian, Discover, Nature, Popular Mechanics, Popular Photography, Popular Science,* and *Field and Stream.* You may obtain back copies of magazines from the li-

ACTIVITY 7.1 • Notes on General Sources

Note Card Format

Reference	Card Number	Call Number
Topic		Location
Points of Interest		
1.		Page
2.		Page
3.		Page
Reasons for Interest		
References Cited		

Tips for Completing Note Cards

1. Write the complete reference using the format designated by your teacher.
2. Record the location of the source; for library material include the call number.
3. Scan the material and identify important information.
4. Record the major topic.
5. Use key words and phrases to record major points. Put quotation marks around the author's own words.
6. Record page numbers; these are needed to accurately document information and to return to sources.
7. Cite interesting references included within the article.
8. Number cards for each source sequentially; place a shortened reference at the top of each card.
9. Use a new card when you change sources.

Relating Library and Scientific Research

1. After you complete all note cards on newspapers and popular journal articles, respond to the following questions:

   ● What **general topic** (*X*) interests you?
   ● What **general action** of (*X*) interests you?

2. After you complete all note cards on general sources, such as textbooks, encyclopedias, dictionaries, respond to the following questions:

   ● What **specific topic** (*X*) interests you?
   ● What **materials** are readily available for you to conduct an experiment on (X)?
   ● What **specific action** of (X) interest you?
   ● How could you **measure** or **describe** the **action** of (X)?

© 2000 by Kendall/Hunt Publishing Company, Cothron, Giese, & Rezba, *Students and Research*.

brarian or request donations from the community. Professional teaching journals such as *The Science Teacher, Science Scope, The American Biology Teacher*, and *The Journal of Chemical Education* contain useful news briefs and descriptions of experiments. File copies of prior student research projects or abstracts can also be a valuable resource.

Next, model two major skills essential for using popular science magazines and newspapers effectively—documenting sources and making note cards. Begin by describing the basic information required to correctly locate and credit major sources of information. In all style manuals, the information cited in Table 7.2 *Documenting Sources* is generally required. For greater relevancy, modify the examples to comply with the style manual used in your school or competitive event, for example, *American Psychological Association, Chicago, Turabian*, and the *Modern Language Association.*

Furnish students with a sample format and instructions for completing a note card on a general source (see Activity 7.1, *Notes on General Sources*). Use a short article related to course content to model how to complete a note card. Emphasize that the purpose of the activity is to obtain ideas for experimentation; voluminous notes are not necessary and will detract from the purpose. Provide additional practice using a variety of newspaper and popular magazine articles.

After taking notes on several newspaper or magazine articles, students should be ready to visit the school library. Arrange for the librarian to show the students the collection of popular magazines and newspapers and to explain procedures for using materials. As an extended assignment, require students to complete note cards on three to five articles of interest. When students have completed the assignment, help them focus on a research topic by responding to two questions. Initially, responses will be very broad and must be narrowed before specific independent and dependent variables can be articulated.

- What general topic (X) interests you?
- What general action of (X) interests you?

General topic	General action
Plants	Growth
Disinfectants	Kills germs
Music	Driving accidents
Glue	Sticking wood
Tobacco	Cellular development
Cold weather	Effectiveness of trumpet oil
Planaria	Regeneration

Stage 2

NARROW THE TOPIC

Textbooks, those handy and overworked encyclopedias of knowledge, represent an underutilized resource for teaching students how to narrow a general topic and action of interest. Library skills essential for documenting, scanning, and

TABLE 7.2   Documenting Sources

Book:	author, title, place of publication, publisher, edition, copyright date, pages
Article:	author, title of article, journal title, volume and issue number, date, pages
Newspaper:	author (if available), title of article, name of paper, date, pages
Abstract:	information needed on the original article is author, title of article, journal title, volume and issue number, date, pages; information needed on the compilation is journal title, volume and issue number, date, pages, abstract number
Unpublished manuscript:	author, title, date; if paper was presented at a meeting add the name, location, and date of the meeting

taking notes from general sources can also be practiced. To maximize class time, use a course-related topic and action to model the process for narrowing a topic. First, make a basic concept map that relates the designated topic (plants) and action (growth). Next, show students how to use their text-

© 1998 PhotoDisc, Inc.

book's index, table of contents, objectives, headings, glossary, terms in italics or bold print, and summary paragraphs to scan for appropriate materials and to identify key terms and phrases. As new terms are found, have the students add them to the concept map as illustrated in Activity 7.2, *Making a Concept Map*. This procedure organizes the student's basic background knowledge and builds a network of descriptors. Knowing many descriptive terms and how they are related will help students effectively use descriptors to locate information in the library.

Use the format for note cards on general sources as a guide to teach students to take notes from their textbooks. Students can readily complete the reference, location, and major topic. Because more information is available in the text than in the popular articles previously used, students may need help in selecting important points. One technique is for students to identify topic sentences and to paraphrase the information. Another method is for students to outline a section of the text. Be patient! Remember, you are simultaneously teaching course content, study skills, and library research skills. If necessary, repeat the activity with other general topics and actions related to course content.

When students are ready to apply skills to their independent research projects, assemble a variety of textbooks in the classroom. Textbook adoption samples can provide a core of elementary, middle, and senior high references in the basic sciences. Professors and local college science departments may also donate textbooks. Because state and regional library accreditation agencies discourage textbooks in the permanent library collection, you will need to establish a textbook

collection within your department and to arrange with the school librarian to put text materials on reserve at specific times of the year.

Have each student draw a basic concept map for his or her topic and action of interest. Help students begin the information search with the easiest available text. Remind students to add descriptors to their concept map and to take succinct notes (see Activities 7.1 and 7.2).

When students have demonstrated the ability to locate and take notes on critical information in textbooks, they are ready to visit the school library. Arrange for the librarian to review or teach students about the card catalog, special sections of the library, and general indexes. Emphasize the use of general and scientific encyclopedias, handbooks, and dictionaries located in the reference section. Monitor students' progress as they begin an extended assignment to complete note cards from three to five general references, such as encyclopedias, dictionaries, and text. Fortified with more knowledge and a well-developed concept map, students can narrow their topic and action of interest by responding to questions.

Q1. What specific topic about $(X \rightarrow \text{plants})$ interests you?
Leaf cuttings

Q2. What materials are readily available for you to conduct an experiment on $(X \rightarrow \text{plants} \rightarrow \text{leaf cuttings})$?
Different varieties of plants, ages of leaves, rooting solutions, light, fertilizer, hormones, soil

Q3. What specific action of $(X \rightarrow \text{plants} \rightarrow \text{growth})$ interests you?
Root propagation

Q4. How could you measure or describe the action of $(X \rightarrow \text{plants} \rightarrow \text{growth} \rightarrow \text{root propagation})$?
Time for roots to appear, number of roots, length of roots, internal development, amount of vascular tissue

ACTIVITY 7.2 • **Making a Concept Map**

Basic Concept Map

1. Respond to the following questions:

   ● What **general topic** (X) interests you?
   RESPONSE   *Plants*

   ● What **general action** of (X ... plants) interest you?
   RESPONSE   *Growth*

2. Draw two circles in the middle of the page. Label the one on the left **topic** and the one on the right **action**.

3. Write the specific topic *(plants)* that interest you inside the left circle and the specific action *(growth)* that interests you in the right circle.

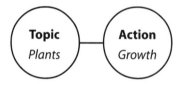

Extended Concept Map

1. Identify scientific terms related to the topic (plant) and action (growth) from general sources, such as textbooks, encyclopedias, and dictionaries. Use the index, table of contents, objectives, headings, glossary, terms in italic or bold print, and summary paragraphs to scan for appropriate materials and to identify key terms and phrases.

2. Add the new terms to the concept map as you find them.

3. Use the new terms to help you organize your background knowledge and to find additional information through library indexes and card catalogs.

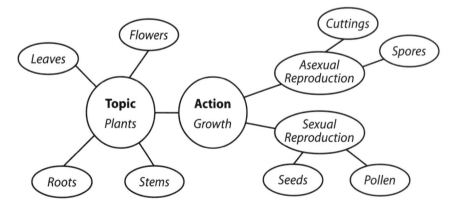

© 2000 by Kendall/Hunt Publishing Company, Cothron, Giese, & Rezba, *Students and Research*.

Establish in your students' minds that the materials available for experimentation on leaf cuttings represent potential classes of independent variables. The methods for measuring or describing the action of interest represent potential dependent variables. With younger students you may end the library research at this point and focus on clarifying the experimental design. For older students, additional library research using scientific journals is recommended before students establish the research topic.

Stage 3

CLARIFY VARIABLES

The focus of library research for older students now changes to scientific research articles related to the potential independent and dependent variables. Because few high school libraries have scientific journals, students are dependent on external library collections. Such library research is inappropriate for students who do not have the scientific knowledge to understand the articles. Because time is critical to both parents and students, it is essential that students know how to use scientific indexes and how to take notes effectively before visiting community libraries.

Prepare students for community libraries by having them read a short report of scientific research related to the course content. Ask students to identify the major sections of the article. Emphasize that most scientific research articles are divided into an introduction, problem statement, methods and materials, and discussion-conclusion. Supply a format for taking notes on scientific research articles (see Activity 7.3, *Notes on Scientific Research*). Have students scan the article to locate the designated information and to record it on the note card. Students may elect to list the components of the experimental design or to draw an experimental design diagram. Because the purpose is to identify major design features, results, and potential areas for research, remind students to keep the notes brief. Have students practice the skills using several different scientific journal articles.

Armed with the ability to take notes, students are ready to learn to access scientific journal articles. Make arrangements for a field trip to a lo-

cal college library and for specific instruction by the librarian on use of the card catalog and major scientific indexes.

Because many high school libraries use the Dewey System and college libraries use the Library of Congress System, assistance with the card catalog is necessary. Allow sufficient time for students to identify several articles appropriate to their research study. Establish an extended assignment that requires students to complete note cards on several scientific journal articles. Be sensitive to students with limited parental support. When students have completed the assignment, help them refine and identify specific independent and dependent variables for investigation by responding to questions.

Q1. What variable will you change to conduct experiments on (X → plants → leaf cuttings)?
Different aged leaves

Q2. What specific changes will you make in (X → plants → leaf cuttings → different aged leaves)?
1-month-old leaves, 2-month-old leaves, 3-month-old leaves

Q3. What action of (X → plants → growth → root propagation) will you investigate?
Growth of roots

Q4. What specific observations or measurements will you make on (X → plants → growth → root propagation → growth of roots)?
Observe the internal development of the xylem and phloem.
Count the number of roots that emerge.

Congratulate students for identifying a specific independent variable to be manipulated, **the age of leaves,** and specific dependent variables to be measured or described, **the number of roots, and the development of the xylem and phloem.**

Stage 4

REFINE THE PROCEDURES

The first three stages led students to the identification of the independent and dependent variables for their investigation. Now they must ar-

ACTIVITY 7.3 • Notes on Scientific Research

Note Card Format

CARD 1

Reference	Card Number	Call Number
		Location
Purpose/Hypothesis:		
Experimental Design:		
Independent Variable:		
Dependent Variable:		
Constants:		
Control:		
Repeated Trials:		

CARD 2

Procedure:
Major Findings/Conclusions:
Areas for Further Research:
References Cited:

Related Library and Scientific Research

After completing all note cards on scientific research articles, respond to the following questions:

- What **variable** will **you change** to conduct an experiment in (X)?
- What **specific change** will you make in (X)?
- What **action** of (X) will you **investigate?**
- What **specific observations** or **measurements** will you make on (X)?

© 2000 by Kendall/Hunt Publishing Company, Cothron, Giese, & Rezba, *Students and Research.*

ticulate the specific materials or organisms to be used, the specific procedures to be followed in conducting the experiment, and the methods by which they will analyze their data. Information can be secured from written materials and by interviewing members of the community.

Excellent sources of information on materials and procedures include K–12 laboratory manuals, handbooks and manuals in specific disciplines, and teacher source books. Initially, students will be screening procedures to identify those that are appropriate, safe, and realistic. Notes should be very brief. If the student later decides to use a particular technique, information can be photocopied.

Students can practice taking notes on technical materials with a class laboratory manual. Prior to a lab, require students to take brief notes using the format illustrated in Activity 7.4, *Notes on Technical Procedures*. Because a class lab is involved, information on material availability and personal assessment of ability to execute will not be as relevant. Strongly emphasize, however, the importance of selecting a procedure that contains readily available and safe materials and that can be conducted correctly and safely by a student.

Community organizations, agencies, and businesses can provide resources and expertise. For example, if a student has decided to conduct an experiment involving African violets, excellent information can be obtained from the agricultural extension service, local greenhouses, or garden clubs. Techniques for observing the development of xylem and phloem can be obtained from specialized handbooks of biological procedures or by interviewing a local university professor. Students can consult telephone directories to locate appropriate community agencies, businesses, organizations, and universities. Frequently, assistance from specific individuals can be obtained by consulting speakers' bureaus provided by professional organizations. Promote good school-community relations by ensuring that students are prepared when they interview community experts. Tips for conducting interviews and recording notes are provided in Activity 7.5, *Conducting an Interview*. Provide students with practice by requiring them to interview students or teachers before visiting community experts.

Stage 5

INTERPRET THE UNEXPECTED

Professional scientists realize that plans and outcomes are not necessarily the same; however, students do not always understand that procedures go awry and unexpected findings occur. These unexpected events represent the nature of all human endeavors, including science, and are a valuable part of the learning process. Whenever unexpected events occur, the preceding stages represent a useful strategy for another round of library research or for narrowing the focus on resources critical to a specific question. Libraries are valuable resources for interpreting such events, modifying procedures, offering explanations for findings, or proposing fruitful avenues for further experimentation.

Because the learning styles of students differ, some are uncomfortable with the linear sequence just proposed. Such students jump among topics, actions, organisms, procedures, and data analysis. Do not despair. The stages can still be useful. Simply provide these students with the overview of the five-stage process illustrated in Table 7.1. As the students focus on a specific component, they can answer the appropriate question. The library skills and resources useful for conducting the library research remain the same at each stage.

MODIFYING YOUR LAB ACTIVITIES

Library research can be easily integrated into the science curriculum. Before students begin a lab activity, identify those parts of the text most directly related to the activity. Assign individual students only short segments of the relevant text to read, usually 1–2 pages. Direct each student to complete a note card on his or her assigned page. You may wish to model this skill by having all students read the same passage and then complete a note card as a whole class lesson. When students have completed their note cards, have them share the main topic and points of interest in their assigned passages.

An alternate lesson involves photocopying selected pages on a particular science topic from a variety of library resources including books,

ACTIVITY 7.4 · Notes on Technical Procedures

Note Card Format

CARD 1

Reference	Card Number	Call Number
		Location
Procedure:		
Major Materials/Equipment:		Availability

CARD 2

Brief Synopsis of Steps:
Your Ability to Implement:
References Cited:

Relating Library and Scientific Research

After completing all the interviews, respond to the following questions:

- What specific **materials** or **organisms** will you use?
- What specific **procedures** will you follow?
- How will you **collect** and **analyze data?**

© 2000 by Kendall/Hunt Publishing Company, Cothron, Giese, & Rezba, *Students and Research.*

ACTIVITY 7.5 • Conducting an Interview

Note Card Format

Reference	Card Number	Phone Number
		Address
Date Arranged:		
Date of Interview:		
Question:		
Response:		

Tips for Conducting Interviews

1. Call the individual at least one week in advance to arrange an interview. Note the time and place carefully. Tell the individual the specific topic to be discussed and information needed.
2. Prepare for the meeting by writing specific questions to be asked. Put each question on a separate card.
3. Listen carefully during the interview. Record important phrases and key words. Tape the interview if the interviewee is agreeable to being taped.
4. Review the notes immediately after the interview and add additional information.
5. Write a letter expressing your appreciation.

Relating Library and Scientific Research

After completing all the interviews, respond to the following questions:

● What specific **materials** or **organisms** will you use?
● What specific **procedures** will you follow?
● How will you **collect** and **analyze data?**

© 2000 by Kendall/Hunt Publishing Company, Cothron, Giese, & Rezba, *Students and Research*.

journals, magazines, and newspapers. Be sure to include relevant reference information. Assign these readings to different students and have them complete note cards. Later, students can share what they learned during a general discussion of what the class as a whole learned about the topic. For a full explanation and several examples of this approach, see Cothron, J.H., Giese, R.N., and Rezba, R.J. (1996). *Science Experiments by the Hundreds.* Dubuque: Kendall/Hunt Publishing Company and the accompanying *Teachers' Guide.*

EVALUATING AND REINFORCING SKILLS

Students' progress with library research can be assessed using a rating sheet, such as *Evaluating for Success 5: Using Library Resources* in Chapter 16, which is subdivided for various types of library materials. Students are awarded 50 points for identification of sources, note taking skills, and organization of information. Criteria for awarding the remaining 50 points differ with the information source—general sources, scientific research articles, manuals of technical procedures, or interviews. These criteria parallel the components of the model note cards and provide additional prompts for the students on procedures for recording information.

REFERENCES

Achtert, W. S., & Gibaldi, J. (1998). *The modern language association style manual.* New York: Modern Language Association of America. (2nd ed.)

The Chicago manual of style: For authors, editors, and copywriters (14th ed.). (1993). Chicago: University of Chicago Press.

Publication manual of the American Psychological Association (4th ed.). (1994). Hyattsville, MD: American Psychological Association.

Turabian, K. L. *A manual for writers of term papers* (6th ed.). (1996). Chicago: University of Chicago Press.

Related Web Sites

http://webster.commnet.edu/mla.htm
http://webster.commnet.edu/apa/apa_index.htm
http://155.43.225.30/workbook.htm
http://www.libertynet.org/lion/lessons.html(lesson plans & activities for teaching the skills of library research using print and electronic mediums)
http://dir.yahoo.com/Education/Instructional_Technology/Online_Teaching_and_Learning/Teacher_Resources/ (Teacher resources for teaching the use of the internet)
http://sunsite.berkeley.eduLibrary Land?chil/skill.htm (Library skills)
http://scout.wisc.edu/scout/indextxt.html (best resources on internet by topic)

CHAPTER 8

Analyzing Experimental Data

Objectives

- Distinguish among quantitative, qualitative, ratio, interval, ordinal, and nominal data; give examples of each.
- Select the appropriate measures of central tendency and variation for a given set of data.
- Describe three ways to find the central value of a set of data, mean, median, mode; compute the values for a set of data.
- Describe two ways to report the variation in a set of data, range, frequency distribution; compute the values for a set of data.
- Construct an appropriate data table and graph for sets of quantitative and qualitative data.
- Teach students the fundamentals of descriptive statistics.
- Use structured procedures to teach students to develop appropriate data tables and graphs for quantitative and qualitative data.

National Standards Connections

- Use appropriate tools and techniques to gather, analyze, and interpret data (NSES).
- Use mathematics in all aspects of scientific inquiry (NSES).
- Understand and apply measures of central tendency and variability (NCTM).

For students to be scientifically literate, they must be able to analyze and interpret data. However, techniques for teaching data analysis and interpretation vary greatly. Some teachers have students answer a series of direct questions about raw data; others require students to mathematically analyze data, to graph data, and to summarize trends. Increasingly, students are required to statistically analyze data in mathematics classes. Despite the emphasis on scientific inquiry in the National Science Education Standards, few science textbooks emphasize statistical analysis of data. Middle and secondary school students can learn to apply basic descriptive statistics to data generated in class activities and in independent projects. In this chapter, strategies are described for teaching students to distinguish among major categories of data, to identify levels of measurement, to compute measures of central tendency and variation, and to construct data tables that communicate descriptive statistics.

REVIEWING BASIC CONCEPTS

Before students can statistically analyze data, they must be able to identify the basic components of an experiment and to draw an experimental design diagram. Use an experimental scenario to review the concepts that were introduced in Chapter 1, *Developing Basic Concepts.*

After students have read the scenario, ask them to identify the independent and dependent variables, constants, control, repeated trials, and hypothesis and to draw an experimental design diagram (see Table 8.1, *Experimental Design Diagram*).

Scenario

Mary investigated the effects of different concentrations of Chemical X on the growth of tomato plants. Mary hypothesized that if higher concentrations of Chemical X were applied, then the plants would exhibit poorer growth. She grew 4 flats of tomato plants, 10 plants/flat, for 15 days. She then applied Chemical X as follows: Flat A: 0% Chemical X; Flat B: 10% Chemical X; Flat C: 20% Chemical X; and Flat D: 30% Chemical X. The plants received the same amount of sunlight and water each day. At the end of 30 days, Mary recorded the height of the plants (cm), the general health of the plants (healthy/unhealthy), and the quality of the leaves using a 4-point scale. Ratings on the leaf quality scale were defined as follows: Rating of 4: Green color, firm, no curled edges; Rating of 3: Yellow-green color, firm, no curled edges; Rating of 2: Yellow color, limp, curled edges; Rating of 1: Brown color, limp, curled leaf.

For additional practice, students may also describe the variables, constants, control, hypotheses, and number of repeated trials in their independent research projects or classroom laboratory experiments. If students have difficulty with these concepts, use strategies described in Chapters 1 and 2 to teach and reinforce these fundamentals.

TYPES OF DATA

In her experiment, Mary recorded three different dependent variables—the height of the plants, the health of the plants (healthy/unhealthy), and the leaf quality (scale of 1–4). Each of the dependent variables can be classified as quantitative or qualitative data.

TABLE 8.1   Experimental Design Diagram

Title: The Effect of Various Concentrations of Chemical X on the Growth of Tomato Plants
Hypothesis: If higher concentrations of Chemical X are applied, then tomato plants will exhibit poorer growth.

IV: Concentration of Chemical X			
0% (Control)	10%	20%	30%
10 plants	10 plants	10 plants	10 plants

DV: Height of plants (cm)
      Health of plants (healthy/unhealthy)
      Leaf quality (scale of 1–4)

C: Amount of sunlight
      Amount of water
      Amount of pre-experiment growth (15 days)
      Length of experiment (30 days)

Quantitative Data

Quantitative data are represented by a number and a unit of measurement that is based upon a standard scale with equal intervals, such as the Metric system of measurement or the Arabic system of numbers. Examples of quantitative variables are the height of a person in meters, the mass of rabbits in kilograms, and the number of seeds that germinated. Quantitative variables may be continuous or discrete. **Continuous quantitative data** are collected using standard measurement scales that are divisible into partial units, for example, distance in kilometers and volume in liters. **Discrete quantitative data** are collected using standard scales in which only whole integers are used, for example, the number of wolves born in a given year or the number of people that can touch their toes. In this chapter, the same statistical techniques will be used with continuous and discrete quantitative data. As shown in the examples below, and explained later (see pages 95–97), graphic presentations differ.

Continuous Quantitative Data

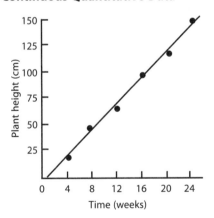

Discrete Quantitative Data

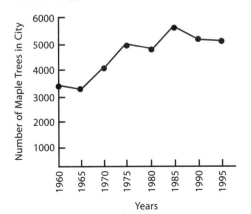

Quantitative data can be further subdivided based on the zero point of the measuring scale. When quantitative data are collected using a standard scale with equal divisible intervals and an absolute zero, it is called **ratio data.** Examples include the temperature of a gas on the Kelvin scale, the velocity of an object in m/sec, and the distance from a point in meters. If the scale does not have an absolute zero, the data are called **interval data.** A common example is the temperature of a substance on the Celsius scale. On this scale, changes in water temperature from 90 to 95 degrees and from 60 to 65 degrees represent the same amount of increase in heat energy or kinetic energy of the molecules, there is no absolute zero, however, because water molecules are still moving at 0°C. In fact, a substance must reach –273°C (0° Kelvin) before molecular motion ceases. In this chapter, the same statistical and graphical techniques are used with ratio and interval data. Mathematically, however, ratio data can be used in a ratio and proportion, whereas, interval data can not. This is why you convert the temperature of gases in degrees Celsius to Kelvin before solving problems with Charles Law or the Perfect Gas Law.

Qualitative Data

Qualitative data are classified into categories. The categories may be discrete categories represented by a word or "number" label or measurements made with a nonstandard scale with unequal intervals. Examples include the gender of an organism and the color of an individual's eyes. The discrete categories are defined by the experimenter and may be based on a literature review or reflect a synthesis of many observations made during experimentation.

Subdivisions of qualitative data are based on the ability to rank order the categories. **Nominal data** exists when objects have been named or placed into discrete categories that cannot be rank ordered, for example, gender (male/female) and the color of hair (red, black, brown). **Ordinal data** exists when objects are placed into categories that can be rank ordered. For example, the activity of an animal could be rated on a scale of 1 to 5, with 5 representing a very active animal. Another example of ordinal data is Moh's Hardness Scale for Minerals. Different statistical techniques and graphic presentations are used for nominal and ordinal data.

Classifying Data

Ask students to classify the three dependent variables in Mary's experiment as quantitative or qualitative data, to identify the level of measurements of each variable, and to justify their answers.

Height of plants
Quantitative (continuous)
Ratio
Equal intervals, absolute zero

Health of plants
Qualitative
Nominal
Discrete categories, not ranked

Leaf Quality
Qualitative
Ordinal
Discrete categories, ranked

DESCRIBING DATA

Statisticians describe a set of data in two general ways. First, they compute a **measure of central tendency** or the one number that is most typical of the entire set of data. Second, they describe the **variation,** or spread within the data. Because the types of scales used to collect quantitative and qualitative data differ, the appropriate measures of central tendency and variation also differ (see Table 8.2).

Measures of Central Tendency

Three different measures of central tendency are available—the mean, median, and mode. The level of measurement of the data determines which measure is appropriate. Definitions and procedures for calculating the mean, median, and mode are given next.

MODE: The value of the variable that occurs most often. It is used for data at the nominal, ordinal, interval, or ratio levels. If two or more values tie in having the most cases, report them as modes.

Examples:

7	15	10
6	13	10
5	12	8
5	11	7
4	9	7
3	9	5
Mode = 5	Mode = 9	4
		Modes = 10 and 7

TABLE 8.2   General Overview of Data Analysis

Type of information	Quantitative data Interval or Ratio	Qualitative data Nominal	Ordinal
What is the most typical or central value?	Mean	Mode	Median
What is the variation or spread?	Range	Frequency distribution	

MEDIAN: The middle value, after all of the cases have been rank ordered from highest to lowest. Half of the cases fall above the median value, half below. The median can be used with ordinal, interval, or ratio data but not with nominal data.

Examples:

7	15	10
6	13	10
5	12	8
5 ☞	11 ☞	7 ☞
4	9	7
3	9	5
Median = 5	Median = 11.5	4
		Median = 7

MEAN: The arithmetic average or the sum of the individual values divided by the number of cases. The mean can only be calculated for interval or ratio data.

$$\text{Mean} = \frac{7 + 6 + 5 + 5 + 4 + 3}{6} = \frac{30}{6} = 5$$

$$\text{Mean} = \frac{15 + 13 + 12 + 11 + 9 + 9}{6} = \frac{69}{6} = 11.5$$

$$\text{Mean} = \frac{10 + 10 + 8 + 7 + 7 + 5 + 4}{7} = \frac{51}{7} = 7.3$$

For ratio and interval data, the mean, median, and mode can be calculated. Because the mean is the most powerful measure of central tendency, it is generally reported for ratio and interval data. The exceptions are those sets of data in which there are a few extreme values that would distort the mean; then the median or mode may be a more accurate measure of central tendency. For ordinal data, both a median and a mode can be calculated. However, the median is generally reported as the more powerful statistic. The mode is the only appropriate measure of central tendency for nominal data.

Measures of Variation

Because students typically compute averages, they are more familiar with the concept of central tendency than the concept of variation within data. Simple measures of variation are the range for a set of quantitative data and the frequency distribution for a set of qualitative data. The **range** is computed by finding the difference between the smallest (minimum) and the largest (maximum) measures of the dependent variable, for example, plant height. Although this value is easy to calculate, students frequently do not comprehend its importance. For example, two experimental groups may have equivalent means yet be very different. Consider John's data on 25 tomato plants grown with a red ground cover and 25 tomato plants grown without a ground cover.

	Red ground cover
Mean Height	15.0 cm
Range in height	10.0 cm
Maximum (largest plant)	18.0 cm
Minimum (smallest plant)	8.0 cm
Number	25 plants

	No ground cover
Mean Height	14.9 cm
Range in height	2.0 cm
Maximum (largest plant)	16.0 cm
Minimum (smallest plant)	14.0 cm
Number	25 plants

Although the means or average heights of the two plant groups were equivalent, the plants grown with the red ground cover exhibited much greater variation.

Frequency Distribution

Similarly, the variation within qualitative data is easier for students to calculate than to conceptualize. The variation is described through a **frequency distribution** that depicts the number of cases falling into each category of the variable, for example, the color of tomatoes produced with different ground covers.

	Red ground cover	
Mode	Pink tomatoes	
Frequency Distribution	Red:	0
	Pink:	12
	Yellow:	8
	Green:	5
Number	25 plants	

	No ground cover	
Mode	Red tomatoes	
Frequency Distribution	Red:	20
	Pink:	5
	Yellow:	0
	Green:	0
Number	25 plants	

Notice how the distribution of colors in the two groups differ. Tomatoes in the red ground cover group are generally pink or yellow, whereas those in the no ground cover group are predominantly red. Advanced students may calculate more powerful measures of variation for quantitative data—the variance and standard deviation. Procedures for calculating these measures are described in Chapters 10 and 11.

At this point, emphasize conceptualization, not calculations, of the two major measures, central tendency and variation. Table 8.2 summarizes the statistical concepts used to describe measures of central tendency and variation in sets of quantitative and qualitative data. Have students use the table to determine the most appropriate measures of central tendency and variation to calculate for each of the dependent variables in Mary's experiment.

Height of plants
Mean
Range

Health of plants
Mode
Frequency distribution

Leaf quality
Median
Frequency distribution

DATA TABLES FOR DESCRIPTIVE STATISTICS

In addition to learning to compute various descriptive statistics, students need to be taught to construct data tables that communicate the statistics. A data table can be constructed easily by combining the rectangular portion of the experimental design diagram with a section listing the specific measures of central tendency, variation, and the number of repeated trials (see Table 8.3 *General Data Table for Descriptive Statistics*, page 91). Review the concept of repeated trials as the number of cases, samples, or individuals who were tested at each level of the independent variable. As with simple data tables, a title that communicates the specific variables being investigated should accompany each table.

DATA TABLE FOR PLANT HEIGHT

Using the raw data on plant height included in Activity 8.1, *Analyzing Experimental Data* (pages 92–93), calculate the mean, range, and number for each concentration of Chemical X. Construct an appropriate data table and enter the computed values (see Table 8.4 *The Effect of Various Concentrations of Chemical X on the Height of Tomato Plants*).

TABLE 8.3   General Data Table for Descriptive Statistics

Title: The Effect of the (IV) on the (DV)			
	Independent variable		
Descriptive information	**Level 1**	**Level 2**	**Level 3**
Central tendency			
Variation		*(Enter computed values)*	
Number			

TABLE 8.4   The Effect of Various Concentrations of Chemical X on the Height of Tomato Plants

	Concentration of Chemical X (%)			
Descriptive information	**0**	**10**	**20**	**30**
Mean	15.3	18.1	10.5	6.0
Range	7.0	6.0	6.0	4.0
Maximum	9.0	20.0	14.0	8.0
Minimum	12.0	14.0	8.0	4.0
Number	10	10	10	10

ACTIVITY 8.1 • Analyzing Experimental Data

Scenario

Mary investigated the effect of different concentrations of Chemical X on the growth of tomato plants. Mary hypothesized that if higher concentrations of Chemical X were applied, then the plants would exhibit poorer growth. She grew 4 flats of tomato plants, 10 plants/flat, for 15 days. She then applied Chemical X as follows: Flat A: 0% Chemical X; Flat B: 10% Chemical X; Flat C: 20% Chemical X; and Flat D: 30% Chemical X. The plants received the same amount of sunlight and water each day. At the end of 30 days, Mary recorded the height of the plants (cm), the general health of the plants (healthy/unhealthy), and the quality of the leaves using a four-point scale. Ratings on the leaf quality scale were defined as follows: Rating of 4: Green color, firm, no curled edges; Rating of 3: Yellow-green color, firm, no curled edges; Rating of 2: Yellow color, limp, curled edges; Rating of 1: Brown color, limp, curled leaf.

Raw Data

Height of plants (cm)				Health of plants				Leaf quality			
Concentration of Chemical X				Concentration of Chemical X				Concentration of Chemical X			
0%	10%	20%	30%	0%	10%	20%	30%	0%	10%	20%	30%
15.0	18.0	12.0	6.0	H	H	H	UN	4	4	2	1
14.0	20.0	10.0	8.0	H	UN	UN	UN	4	3	3	1
13.0	14.0	14.0	5.0	H	H	H	UN	4	3	3	1
15.0	20.0	10.0	4.0	H	H	UN	H	4	4	2	2
15.0	18.0	8.0	4.0	H	H	H	UN	4	4	2	1
17.0	19.0	8.0	5.0	H	UN	UN	UN	4	2	2	1
18.0	18.0	10.0	8.0	H	H	UN	H	4	4	2	2
12.0	18.0	10.0	7.0	H	H	UN	UN	4	4	3	2
19.0	17.0	11.0	8.0	H	H	UN	UN	4	4	2	1
15.0	19.0	12.0	5.0	H	H	H	UN	4	3	2	1

(H = Healthy; UN = Unhealthy)

Data Analysis

Directions

1. Read the scenario of Mary's experiment and identify the independent variable, dependent variables, constants, control, hypothesis, and repeated trials. Draw an experimental design diagram.
2. Classify each of the dependent variables in Mary's experiment as quantitative or qualitative data; justify your answer.

(continued on the following page)

ACTIVITY 8.1 • Analyzing Experimental Data *(continued)*

3. Classify each of the dependent variables in Mary's experiment as nominal, ordinal, interval, or ratio data; justify your answer.
4. For each of the dependent variables in Mary's experiment, describe the most appropriate measures of central tendency and variation.
5. Construct a data table for displaying the measures of central tendency, variation, and number for each set of raw data. Compute the appropriate measures of central tendency and variation, and enter them into the table. Construct an appropriate graph.

USING TECHNOLOGY ·

1. In the **STAT** mode of your calculator, enter the values for heights of plants for 0% in List 1, for 10% in List 2, for 20% in List 3, and for 30% in List 4. (See Appendix A, *Using Technology,* for additional help in using the graphing calculator).
2. In the **STAT** mode select CALC (for calculate) and then 1 VAR (for 1-variable statistics). Depending on the brand of your calculator, you will need to enter the desired list number (e.g., L1), or SET the 1-variable x-list to the desired list number before selecting 1 VAR. *Repeat* the selection process for each set of data by changing the list number.
3. Among the calculated values provided are the mean and median for the data set. Calculators that also provide a value for the mode will display the highest mode if there is more than one mode for that set of data. In addition, the range can be calculated from the maximum and minimum values given for each data set.
4. Use appropriate values to construct your summary data table and bar graph.
5. Now, repeat these steps for the leaf quality data. In the **STAT** mode of your calculator, enter the rating values (1, 2, 3, 4) for the various chemical concentrations in Lists 1–4. Sort each list in ascending order and identify the median value for each chemical concentration.
6. Use the appropriate values to construct your summary data table and frequency distribution.

© 2000 by Kendall/Hunt Publishing Company, Cothron, Giese, & Rezba, *Students and Research.*

The selection and construction of an appropriate type of graph for given sets of data are described in Chapter 5, *Constructing Tables and Graphs.* If students have not mastered these concepts, use the activities suggested in that chapter to reinforce the fundamentals. Because both the independent variable, concentration of Chemical X, and the dependent variable, plant height (cm) are quantitative continuous data, either a bar or line graph can be used. Because the levels of the independent variable are few and widely separated, a bar graph was selected, as illustrated in Figure 8.1.

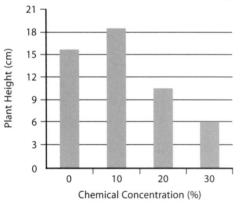

Figure 8.1 The Effect of Various Concentrations of Chemical X on Plant Height (cm).

Data Table for Plant Health

Because the health of the plant was reported as qualitative data at the nominal level of measurement, the mode is the appropriate measure of central tendency. Variation is reported through a frequency distribution. Use the data from Mary's experiment to construct an appropriate data table for summarizing plant health at each concentration of Chemical X (see Table 8.5 *The Effect of Various Concentrations of Chemical X on the Health of Tomato Plants*).

TABLE 8.5   The Effect of Various Concentrations of Chemical X on the Health of Tomato Plants

Descriptive information	Concentration of Chemical X (%)			
	0	10	20	30
Mode	Healthy	Healthy	Unhealthy	Unhealthy
Frequency distribution				
Healthy	10	8	4	2
Unhealthy	0	2	6	8
Number	10	10	10	10

Because the dependent variable is qualitative (nominal), a special type of bar graph—a frequency distribution—is used to graph the data. A frequency distribution shows the number of plants that fall into each category, healthy and unhealthy. On the graph, place the independent variable, concentration of the chemical, on the X axis. Subdivide the Y axis into an appropriate scale for plotting the number of plants that fall into each category of the dependent variable, plant health. Graph the frequency distribution by drawing a vertical bar from a specific concentration of Chemical X to the appropriate number of plants falling into each category, healthy and unhealthy. As illustrated in Figure 8.2, use a key, such as white and shaded bars, to depict each category. From the graph, both the frequency distribution and mode for each concentration of Chemical X can be determined.

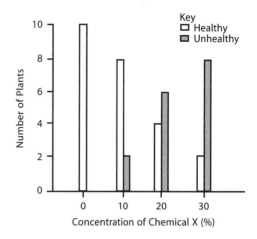

Figure 8.2 Frequency Distribution for Plant Health at Various Concentrations of Chemical X.

DATA TABLE FOR LEAF QUALITY

Mary used a 4-point scale for rating leaf quality. Because the scale involved nonstandard measurements and placement into categories that could be ordered, the median is the appropriate measure of central tendency. Variation is reported through a frequency distribution. Use the raw data on leaf quality from Activity 8.1 to compute the median and frequency distribution (see Table 8.6 *The Effect of Various Concentrations of Chemical X on Leaf Quality*).

Leaf quality represents qualitative (ordinal) data; thus, a bar graph is appropriate. Again, subdivide the X axis to depict the various concentrations of the chemical. Use the 4-point rating scale for leaf quality as subdivisions of the Y axis. Graph the median leaf quality value for each concentration of Chemical X (see Figure 8.3). Using procedures previously described, the frequency distribution for the number of plants falling into each category of leaf quality may also be graphed (see Figure 8.4).

TABLE 8.6   The Effect of Various Concentrations of Chemical X on Leaf Quality

Descriptive information	Concentration of Chemical X (%)			
	0	10	20	30
Median	4	4	2	1
Frequency distribution				
Quality 4	10	6	0	0
Quality 3	0	3	3	0
Quality 2	0	1	7	3
Quality 1	0	0	0	7
Number	10	10	10	10

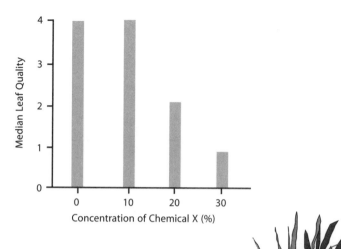

Figure 8.3 Median Leaf Quality for Plants Exposed to Various Concentrations of Chemical X.

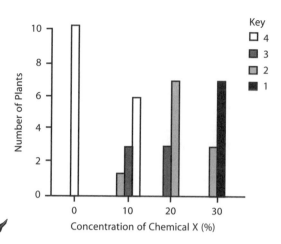

Figure 8.4 Frequency Distribution of Leaf Quality for Plants Exposed to Various Concentrations of Chemical X.

MAKING DECISIONS ABOUT DESCRIPTIVE STATISTICS AND GRAPHS

At this point, students should be able to statistically describe their data and to determine an appropriate measure of central tendency and dispersion/variation for a specific experiment. How do students decide upon a graphic presentation? Using Table 8.7, *Making Decisions about Descriptive Statistics & Graphs,* ask students to determine if the levels of the independent variable are continuous or discrete. Then, ask them if the dependent variable is quantitative (continuous/discrete), qualitative (ordinal), or qualitative (nominal).

1. *Are the levels of the independent variable continuous or discrete?*
   Continuous means that the levels of the variable are not separate categories and that the intervals between the levels have meaning. An example is the sugar content of a fluid expressed as 0%, 5%, 10%, 15%, 20%, 25%, and 30%. Discrete, however, means that the categories are separate and not continuous. An example of discrete levels of a variable is the sugar content of fluid expressed as low, medium, and high concentration.

2. *Is the dependent variable quantitative (continuous/discrete), qualitative (ordinal), or qualitative (nominal)?*

If the dependent variable is **quantitative (continuous/discrete)** then three measures of central tendency can be used: the mode, median, or mean. Generally, the mean is used when the data follows a normal distribution (mound shape). When data are skewed, the median may be a more appropriate measure of central tendency. The range is an appropriate measure of dispersion/variation.

If the independent variable is discrete, then bar graphs are the only option for displaying the data. If the independent variable is continuous, either bar or line graphs may be used. Line graphs, with a line-of-best-fit, are used with continuous quantitative data. Broken line graphs, in which the data points are connected, are used with discrete quantitative data. The same statistical and graphical presentations are used for ratio and interval data.

If the dependent variable is **qualitative (ordinal),** then two measures of central tendency can be used, i.e., the mode and median. Generally, the median is used. A frequency distribution is used to communicate dispersion/variation within the data. For the graphical presentation, you may make a bar graph of the medians or a frequency distribution.

If the dependent variable is **qualitative (nominal),** then the measure of central tendency is the mode and the measure of dis-person/variation is the frequency distribution, which is also used for the graphical presentation.

Have students read Investigation 8.1 *A Sudsy Experience,* (see page 100). Then, use the information in Table 8.7 to help students make decisions about appropriate descriptive statistics and graphical displays for the experiment. Students should conduct the experiment and display the data through tables and appropriate graphs.

MODIFYING YOUR LAB ACTIVITIES

Lab activities that characterize the nature of science are those that require qualitative and quantitative measurements and the accurate collection, display, and analysis of data. When these are lacking in activities, modifications can be made to introduce new concepts or to provide additional practice opportunities for students. For example, the activity illustrated in Modified Example 8.1, *Pupillary Reflex,* can be quantified by giving students the following directions:

- Place a mm ruler across the bridge of your partner's nose.
- Measure the size of the pupil under each set of light conditions.
- Repeat five times to increase the reliability of your results.

Human error in making scientific measurements creates unavoidable uncertainty. Because errors are as likely to be larger and smaller than the true value, the mean (average) of many measurements will be more reliable than just one. Beginning data analysis skills should be emphasized by asking students to find the mean pupil size for each set of light intensities, dark, medium and high.

TABLE 8.7   Making Decisions about Descriptive Statistics & Graphs

Levels of Independent Variable	Dependent Variable	Central Tendency	Dispersion/ Variation	Graphic Displays
Continuous	Quantitative (continuous/ discrete)	Mode Median Mean	Range	Bar Graph Line Graph, Line-of-Best Fit (continuous) Line Graph, Broken Line (discrete)
	Qualitative (ordinal)	Mode Median	Frequency distribution	Bar Graph (medians) Frequency distribution
	Qualitative (nominal)	Mode	Frequency distribution	Frequency distribution
Discrete	Quantitative (continuous/ discrete)	Mode Median Mean	Range	Bar Graph
	Qualitative (ordinal)	Mode Median	Frequency distribution	Bar Graph (medians) Frequency distribution
	Qualitative (nominal)	Mode	Frequency distribution	Frequency distribution

MODIFIED EXAMPLE 8.1 • Pupillary Reflex

Purpose: To observe an inborn reflex.

Materials: Medium intensity light source (window or overhead light)
High intensity light source (flashlight)
Pair of students

Procedure

In conducting this experiment, you will need to work in pairs. Have your partner **close and cover both eyes** for 30 seconds and then open them quickly. Immediately observe your partner's pupils. Record your observations. Have your lab partner look at a **medium intensity light source,** such as a window or overhead light for 30  seconds. Observe your partner's pupils. Record your observations. Repeat the procedure with your partner looking at a **high intensity light source,** such as a flashlight, for 30 seconds. Again, observe your partner's pupils and record the observations.

Analysis

1. Which part of the eye changed?
2. How quickly does the change occur?

Modifications for Analyzing and Communicating Experimental Data

1. Measure the size of the pupil under each set of light conditions (dark, medium, high) by placing a millimeter (mm) ruler across the bridge of your partner's nose. Repeat five times to increase the reliability of your results.
2. What are the appropriate measures of central tendency and variation to use in analyzing the data?
3. Construct a data table for displaying descriptive statistics for the class data. Compute the values and enter them into the table.
4. Write a paragraph of results about the data.
5. Write a conclusion for the experiment.

TABLE 8.8 Data Table for Descriptive Statistics on Pupillary Reflex

Descriptive Information	Light Intensity		
	Dark	**Medium**	**High**
Mean			
Range			
Maximum			
Minimum			
Number			

To introduce or reinforce descriptive statistics, students could be asked to complete a summary data table, identifying the mean, range, and maximum and minimum values for each light condition (see Table 8.8 *Data Table for Descriptive Statistics on Pupillary Reflex*).

EVALUATING AND REINFORCING SKILLS

For students to fully conceptualize measures of central tendency and variation, numerous experiences with a variety of data are essential. As previously discussed, modifying classroom laboratory activities to incorporate descriptive statistics can provide these opportunities. Other examples of experimental scenarios and representative methods of data analysis are included in Chapter 10, *Displaying Dispersion/Variation within Data* and Chapter 11, *Determining Statistical Significance.* For mathematical explanations of the statistical concepts, consult the chapter references.

Evaluation suggestions for paper-pencil tests are included in Chapter 15. Items require students to classify data, to identify appropriate measures of central tendency and variation, to calculate various descriptive statistics, and to make judgments about the most appropriate statistics for a given set of data. In Chapter 16, the first parts of *Evaluating for Success 7: Analyzing and Communicating Data: Description Statistics* can be used as a rubric to teach the skills or as an assessment tool.

REFERENCES

Landwehr, J.M. & Watkins, A.E. (1994). *Exploring data*. A component of the Quantitative Literacy Series. Palo Alto, CA: Dale Seymour Publications.

Landwehr, J.M., Swift, J., & Watkins, A.E. (1998). *Exploring surveys and information from samples.* A component of the Data-driven Mathematics Series. Palo Alto, CA: Dale Seymour Publications.

McClave, J.T., Dietrich, Frank H. II, & Sincich, Terry. (1997). *Statistics* (7th ed.) Upper Saddle River, NJ: Prentice-Hall, Inc.

Yates, D.S., Moore, O.S., McCabe, G.P. (1999). *The Practice of Statistics: TI-83 Graphing Calculator Enhanced.* New York: W.H. Freeman and Company.

Related Web Sites

http://www.scri.fsu.edu/~dennisl/CMS/sf/sf_details.html (standard deviation)
http://www.mste.uiuc.edu/stat/stat.html

INVESTIGATION 8.1 • A Sudsy Experience

Question

How do minerals in water affect its ability to form suds?

Materials for Each Group of Two Students

3	100 ml graduated cylinders
3	pieces of plastic wrap (25 cm²) torn from roll
3	rubber bands (optional)
50 ml	distilled water
50 ml	water with borax (2 g of borax per 1000 ml of distilled water)
50 ml	water with Epsom salt (2 g of Epsom salt per 1000 ml of water)
1	metric balance
1	bar of Ivory soap
1	device to scrape soap such as plastic knife, metal spoon, or vegetable peeler
1	small container, such as a plate or evaporating dish for scraped soap
1	roll of paper towels or newspapers
1	metric ruler
1	masking tape for labels

Safety

- Wear goggles.
- Wash hands.
- Dispose of chemicals in marked containers.

Procedure

1. Place newspaper or paper towels on work surface in case of spills.
2. Put 50 ml of distilled water in a 100 ml graduated cylinder. Use a piece of masking tape to label the container. Add 1 g of "shaved" Ivory bar soap.
3. Place the plastic wrap over the top of the cylinder, being careful to seal well. To help insure a tight seal, you may also use a rubber band to secure the plastic wrap. Hold the cylinder straight in front of your body. Place one hand at the top of the cylinder and the other where the plastic wrap ends. Have your partner hold a Metric ruler in front of you; use the length of the ruler as the "length of a shake." Shake the container up and down 10 times.
4. Remove the plastic wrap. Measure the height of the suds column in millimeters (mm). Record the measurement.
5. Place the cylinder in the middle of the work surface. Be sure that you have labeled it so that you will know the type of water inside. After 30 minutes, classify the type of suds formed using the following categories.

(continued on the following page)

INVESTIGATION 8.1 • A Sudsy Experience (continued)

LClN	Light foamy suds, cloudy water, no large soap particles
LClF	Light foamy suds, cloudy water, few large soap particles
TClF	Thick suds, cloudy water, few large soap particles
TCM	Thick suds, clear water, many large soap particles

6. Repeat Steps 2 to 5 using 50 ml of water with borax and 50 ml of water with Epsom salt.
7. Enter your data in a class data table. Be sure that at least 10 sets of data are entered. If there are fewer than 10 student groups, then repeat Steps 1–6 to obtain sufficient trials.

Class Data Table I: Height of Suds Column (mm)

	Height of column (mm)													
	Trials (lab work groups)													
Type of water	1	2	3	4	5	6	7	8	9	10	11	12	13	etc.
Distilled water														
Water with borax														
Water with Epsom salt														

Class Data Table II: Type of Suds Formed (LClN, LClF, TClF, TCM)

	Type of suds formed													
	Trials (lab work groups)													
Type of water	1	2	3	4	5	6	7	8	9	10	11	12	13	etc.
Distilled water														
Water with borax														
Water with Epsom salt														

Analyzing Your Data

1. Make a summary data table and graph to display the class data on "Height of Suds."
2. Make a summary data table and graph to display the class data on "Type of Suds Formed."

(continued on the following page)

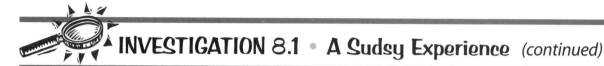

INVESTIGATION 8.1 • A Sudsy Experience *(continued)*

USING TECHNOLOGY ·

1. In the **STAT** mode of your calculator, enter the values for heights of soap suds for distilled water in List 1, for water with borax in List 2, and for water with Epsom salts in List 3. (See Appendix A, *Using Technology*, for additional help in using the graphing calculator).

2. In the **STAT** mode select CALC (for calculate) and then 1 VAR (for 1-variable statistics). Depending on the brand of your calculator, you will need to enter the desired list number (e.g., L1), or SET the 1-variable x-list to the desired list number before selecting 1 VAR. *Repeat* the selection process for each set of data by changing the list number.

3. Among the calculated values provided are the mean and median for the data set. Calculators that also provide a value for the mode will display the highest mode if there is more than one mode for that set of data. In addition, the range can be calculated from the maximum and minimum values given for each data set.

4. Use appropriate values to construct your summary data table and bar graph.

Extending Your Learning

1. How does Ivory bar soap differ from other brands of hand soap? Would you get the same results with other types of soap? Why?
2. How did the presence of borax and Epsom salt affect the suds column? Did the data support your hypothesis?
3. What is the chemical formula for Epsom salts?
4. Why is water containing magnesium and calcium ions called "hard water"? How does the presence of these ions affect suds formation? Cleaning?
5. What is the chemical formula for borax? How does the presence of borax affect suds formation? Cleaning?
6. How does home water softening unit work? What types exist?
7. Why are some types of water softening units not recommended for people with high blood pressure, heart disease, and kidney disease?
8. How could you improve this experiment?
9. What other experiments might you conduct on this topic?

© 2000 by Kendall/Hunt Publishing Company, Cothron, Giese, & Rezba, *Students and Research*.

Practice

1. For *each* of the experimental scenarios below, state the following information.

    A. Type of data (quantitative or qualitative) collected for the dependent variables.
    B. Level of measurement (nominal, ordinal, interval, or ratio) represented by the measures of the dependent variables.
    C. Most appropriate measure of central tendency for describing the dependent variable.
    D. Most appropriate measure of variation for describing the dependent variable.
    E. Construct a data table for the data. Compute the appropriate measures of central tendency, variation, and number; enter them in the table.
    F. Construct an appropriate graph.

The heated soil scenario: Walter placed 1 cup of sand (S), potting soil (P), and a mixture of sand and potting soil (M) into separate pint-size containers. In each of the containers he placed a thermometer so that the bulb was 2.5 cm below the surface. He placed the 3 containers under identical heat lamps for an hour. The original temperature of each jar was 15°C. After heating the jars the first time, the temperatures of the containers were S = 28°C, P = 33°C, and M = 29°C. After heating the jars a second time, the temperatures of the contents were S = 26°C, P = 29°C, M = 29°C. After the third heating, the temperatures were S = 27°C, P = 31°C, M = 22.5°C. Between each heating, the contents of the jars were cooled to 15°C.

The peat moss scenario: Norm wanted to know if adding peat moss to sand would affect its ability to hold water. He put 200 ml of pure sand into container A. He put a mixture of 80% sand and 20% peat moss into container B. Into container C he placed a mixture of 60% sand and 40% peat moss. Finally, he placed a mixture of 40% sand and 60% peat moss into container D. He added water to each container and measured the amount of water the contents would absorb. He dried the sand and peat moss. He repeated the experiment 5 times. He collected the following data.

Composition of mixture	Water holding capacity (ml)				
	Trial 1	Trial 2	Trial 3	Trial 4	Trial 5
100% sand	74	80	70	71	74
60% sand; 40% peat moss	86	88	90	92	94
40% sand; 60% peat moss	110	116	104	108	112
80% sand; 20% peat moss	84	82	86	82	84

(continued on the following page)

Practice *(continued)*

2. Use the following experimental situation to answer Questions 2A–2D:

   John raised deer mice. From one litter, he obtained mice with the following masses:

Mouse	Mass (g)
A	1.5
B	2.5
C	3.0
D	2.5
E	2.0
F	1.5
G	2.5

   A. Which type of data is represented (qualitative or quantitative)?
   B. What is the minimum value of the masses of the mice?
   C. What is the range of mouse mass?
   D. Compute the mean, mode, and median.

3. Use the following experimental situation to answer Questions 3A–3C. John sampled the apples on 5 trees in three different orchards, with different varieties, to determine the stage of ripeness. For Variety 1, he counted 70 dark green, 60 yellow-green, 80 pink and 60 red apples. For Variety 2, he counted 40 dark green, 120 yellow-green, 50 pink and 60 red apples. For Variety 3, he counted 10 dark green, 30 yellow-green, 45 pink and 185 red apples.

   A. Which level of measurement does the data represent, e.g. nominal, ordinal, interval?
   B. Construct an appropriate data table.
   C. Construct an appropriate graph.

© 2000 by Kendall/Hunt Publishing Company, Cothron, Giese, & Rezba, *Students and Research.*

············· C H A P T E R 9 ·············

Communicating Descriptive Statistics

Objectives

- Write appropriate paragraphs of results for sets of quantitative and qualitative data.
- Write an appropriate conclusion for an investigation.
- Use structured procedures to teach students to write paragraphs about quantitative and qualitative data.
- Use a structured procedure to teach students to write a conclusion for an investigation.

National Standards Connections

- Recognize and analyze alternative explanations and predictions (NSES).
- Think critically and logically to make the relationships between evidence and explanations (NSES).
- Evaluate arguments that are based on data analysis (NCTM).

Remember the last time you asked your students to write something? Whether that something was a report, an introduction to a research project, a paragraph explaining a graph, or a conclusion to a laboratory report, what were the typical student comments? "I don't know where to start." "What do I need to look up?" "I know what it means, but I just can't say it." "What's a conclusion?" These and similar comments are common. They reflect the difficulty students have in moving from unfocused thoughts to precise written language.

Because writing about experimental results involves students in both analyzing and synthe-sizing relationships, a framework for guiding writing is frequently needed. Structured guidelines for writing about descriptive statistics in a data table can provide such a framework. Descriptive statistics include two major types of information—a measure of central tendency and a measure of variation. The terms statisticians use to describe the measures of central tendency and variation in sets of quantitative and qualitative data are summarized in Table 9.1 *Describing Quantitative and Qualitative Data*. For a more detailed explanation, see Chapter 8, *Analyzing Experimental Data*.

TABLE 9.1   Describing Quantitative and Qualitative Data

Type of data	Central or typical value	Variation or spread in data
Quantitative data: data based on measurements or counts with a standard scale	Mean	Range
Qualitative data: data based on descriptive observations using a nonstandard scale	Median Mode	Frequency distribution

This chapter presents suggestions for writing paragraphs about quantitative and qualitative data and conclusions (see Table 9.2 *How to Communicate Descriptive Statistics*). Data tables presented in Chapter 8 form the basis for the writing activities. The tables depict the effects of various concentrations of Chemical X on the height, health, and leaf quality of tomato plants.

WRITING ABOUT QUANTITATIVE DATA

Provide students with a data table containing descriptive statistics on quantitative data such as tomato plant heights at various concentrations of Chemical X. (See Table 9.3 *The Effect of Various Concentrations of Chemical X on Height of Tomato Plants.*)

Review the meaning of the terms mean, range, and number. Discuss patterns in the data including changes in the mean height and the range with increasing concentrations of Chemical X. This discussion is very important and forms the basis on which students will begin to write a paragraph using a four-step process.

Step 1:   Write a *topic sentence* stating the independent and dependent variables, and a reference to tables and graphs.

Step 2:   Write sentences comparing the *measures of central tendency* of the groups.

Step 3:   Write sentences describing the *variation* within the groups.

Step 4:   Write sentences stating how the data *support the hypothesis.*

Step 1: Write Topic Sentence

To begin their paragraphs, have students write a topic sentence about the data table. The topic sentence should state the independent and dependent variables and make reference to appropriate tables and graphs. Before students write sentences, provide several examples, such as "The effect of various amounts of sugar on the activity of mice is depicted in Table X," or "Table Y depicts the effect of various types of water pollutants on the walking speed of a water strider." Allow a few minutes for students to write their topic sentences. Divide the blackboard into a section for steps and two or three sections for examples (see Table 9.4 *Writing about Quantitative Data*). Select several students who were successful but who have different styles of sentences. Ask them to write their topic sentences on the board. Emphasize that there are many ways to write a topic sentence as long as all the needed information is communicated. In general, students find it easy to write the topic sentence. Praise the high success rate and thereby establish a positive climate for the more involved steps that follow.

Step 2: Compare Measures of Central Tendency

Next, have students write one or more sentences that compare the means or typical values of the groups. Again, provide several examples from other sets of data: "The mean mass (150 g) of Elodea grown under blue light was higher than the mean mass (50 g) of the control," or "Water striders have a mean speed of 5 cm/min on unpolluted water as compared with mean speeds of 4 cm/min

TABLE 9.2 How to Communicate Descriptive Statistics

Results section of a scientific paper
Quantitative Data: *Write a paragraph that includes a topic sentence, comparison of the means, description of the variation, and support for the hypothesis. Use four steps to guide your writing:*   1. Write a topic sentence stating the independent and dependent variables, and a reference to tables and graphs.   2. Write sentences comparing the measures of central tendency (means) of the groups.   3. Write sentences describing the variation within the groups.   4. Write sentences stating support of the hypothesis by the data.
Qualitative Data: *Determine the type of qualitative data, nominal or ordinal. Then write a paragraph that includes a topic sentence, comparison of the measures of central tendency, description of the variation, and support for the hypothesis. Again, use four steps to guide your writing:*   1. Write a topic sentence stating the independent and dependent variables, and a reference to tables and graphs.   2. Write sentences comparing the measures of central tendency (mode or median) of the groups.   3. Write sentences describing the variation (frequency distribution) within the groups.   4. Write sentences stating how the data support the hypothesis.
Conclusion section of a scientific paper
Describe the purpose, major findings, an explanation for the findings, and recommendations for further study. Use six questions to guide how you write the conclusion: 1. What was the purpose of the experiment?   2. What were the major findings?   3. Was the hypothesis supported by the data?   4. How did your findings compare with other researchers?   5. What possible explanations can you offer for the findings?   6. What recommendations do you have for further study and for improving the experiment?

© 2000 by Kendall/Hunt Publishing Company, Cothron, Giese, & Rezba, *Students and Research.*

TABLE 9.3 The Effect of Various Concentrations of Chemical X on Height of Tomato Plants

Descriptive information	Concentration of Chemical X			
	0%	**10%**	**20%**	**30%**
Mean	15.3 cm	18.1 cm	10.5 cm	6.0 cm
Range	7.0	6.0	6.0	4.0
Maximum	19.0	20.0	14.0	8.0
Minimum	12.0	14.0	8.0	4.0
Number	10	10	10	10

TABLE 9.4 • Writing about Quantitative Data

Steps	Examples	Examples
1. Write a **topic sentence** stating the independent and dependent variables, and a reference to tables and graphs.	The effect of various concentrations of Chemical X on the height of tomato plants is summarized in Table 9.3.	Table 9.3 shows the effect of various concentrations of Chemical X on the growth of tomato plants.
2. Write sentences comparing the **measures of central tendency (means)** of the groups.	The mean height of plants grown at 10% X (18.1 cm) was higher than the control (15.3 cm). At higher concentrations of Chemical X, mean plant height was reduced, e.g., 10.5 cm at 20% X and 6.0 cm at 30% X.	The mean height of plants decreased as the concentration of Chemical X increased, e.g., 18.1 cm, 10.5 cm, 6.0 cm. Interestingly, greater mean height was observed at 10% X (18.1 cm) than in the control group (15.3).
3. Write sentences describing the **variation (range or standard deviation)** within the groups.	The range in plant height in the control was greater than the groups receiving 10% X, 20% X and 30% X. As the concentration of Chemical X increased, the range in plant height decreased.	The control group, 10% X group, and 20% X group showed similar variations in height. Less variation occurred at 30% X (Range = 4 cm).
4. Write sentences stating how the data **support the hypothesis.**	The data partially supported the hypothesis that plant growth would decrease as concentration of Chemical X increased. Below 10% X little growth differences were observed; however, at higher concentrations, growth was severely retarded.	At 10% X, growth was not retarded. Above 10% X, plant growth was progressively retarded; thus, the hypothesis was partially supported that growth of plants would decrease as concentrations of Chemical X increased.

© 2000 by Kendall/Hunt Publishing Company, Cothron, Giese, & Rezba, *Students and Research.*

on water polluted with gasoline and 2 cm/min on water polluted with diesel fuel." Allow students time to write their sentences. Monitor progress closely, prompting students on an individual basis. As with the topic sentence exercise, have a few students who were successful with this task share their sentences. Write several different, but correct, examples on the board. After the examples have been presented, give students an opportunity to revise their sentences as necessary. If students experience difficulty, provide additional practice using other data tables before they proceed to the next step.

Be prepared for great variation in students' abilities to analyze and write about trends within the data. Because of their concrete cognitive level, many middle school students will just report the data rather than write summary sentences about it. For instance, they may report that the mean plant heights were 15.3 cm, 18.1 cm, 10.5 cm, and 6.0 cm rather than state that the mean plant height decreased as the concentration of Chemical X increased. Accept the students' first efforts as a beginning. By practicing writing techniques throughout the year, students' writing will improve.

Step 3: Describe Variation Within Data

Discuss with students the concept that variation within a set of data is important and should be reported. A group of plants that vary in size from 2 to 12 cm is different from a group of plants that vary in size from 7 to 9 cm, even if the two groups have identical means of 8 cm. The first group has a large range (10 cm) or variation in plant height, while the second group has a small range (2 cm) with little variation in plant height. Ask students to write one or more sentences that describe the ranges within the groups. As before, provide several examples: "Rats exposed to high levels of sugar had a small range of .5 turns/second, while those in the control and low sugar groups exhibited a large range of 8 turns/second," or "Plants in the experimental and control groups exhibited similar variations in size, ranging from 15 to 18 cm in height." Generally, students will proceed faster with these sentences than with the sentences comparing means. Again, have students with different but acceptable sentences share them with the class, adding several good examples to the board as in Steps 1 and 2.

Step 4: State Support for Hypothesis

Conclude the paragraph with a sentence stating how the data support the hypotheses such as: "Because the mean height of plants receiving fertilizer was greater than the control group, the hypothesis that plant growth would be increased by fertilizer was fully supported," or "Because the mean heights of plants receiving various amounts of compost were equivalent to the control group, the hypothesis that . . . was not fully supported." Again, put several acceptable sentences on the board.

Construct a paragraph by combining the sentences that resulted from completing the four steps. Ask students to react to the overall meaning of the paragraph. If the meaning is clear, improve the paragraph by using more effective transition phrases, maintaining the same tense, and so on. As a final step, edit the paragraph for grammar and spelling.

Using another data table, repeat the writing steps as a whole-class activity. Then, give students a new data table and have them write a paragraph in small groups or as a homework assignment.

WRITING ABOUT QUALITATIVE DATA

The same process used to teach the writing of a paragraph about quantitative data can be used to teach students to write paragraphs about qualitative data. Provide students with a data table containing descriptive statistics on qualitative data such as the effect of Chemical X on the health and leaf quality of tomato plants (see Tables 9.5 *Writing about Nominal Data* and 9.6 *Writing about Ordinal Data*). Review the distinguishing characteristics of nominal and ordinal data and the appropriate measures of central tendency and

Photograph by Jeannette Bishop; print by Eleanor and Wilton Tenney.

variation for each type of data. After discussing the influence of Chemical X on plant health, have students use the same four-step procedure to write a paragraph. Simply modify the questions to reflect the appropriate measures of central tendency (mode or median) and variation (frequency distribution).

Step 1: Write a *topic sentence* stating the independent and dependent variables, and a reference to tables and graphs.

Step 2: Write sentences comparing the *measures of central tendency* (modes or medians) of the groups.

Step 3: Write sentences describing the *variation (frequency distribution)* within the groups.

Step 4: Write sentences stating how the data *support the hypothesis.*

Nominal Data

Following the same procedure used with quantitative data, guide the students sentence by sentence through writing a paragraph that describes the influence of Chemical X on the health of tomato plants. An illustrative data table and examples of student sentences are provided in Table 9.5.

Ordinal Data

Use the same process to write a paragraph of results about the influence of Chemical X on leaf quality. Because leaf quality is ordinal data, the medians of the group are compared. As with the data on plant health, the variation in leaf quality is described by the frequency distributions. Examples of students' sentences are provided in Table 9.6.

WRITING A CONCLUSION

A similar guided approach for helping students write a conclusion for a scientific investigation is described in Chapter 6, *Writing Simple Reports.* Questions that have proven helpful to students and illustrative sentences based on the data in Tables 9.3, 9.4, and 9.5 are described in Table 9.7 *Writing a Conclusion.*

The writing techniques presented in this chapter are very structured. They do provide, however, a necessary organizational framework that enhances the students' initial success and promotes a positive attitude toward writing in science. As students become more proficient with technical writing, you will see individual styles emerge that communicate all the necessary information.

Suggestions for integrating technical writing into the science curriculum are provided in Chapter 6, *Writing a Simple Report.* Procedures for writing formal papers, such as those required for competitions, are described in Chapter 13, *Writing Formal Papers.* Techniques for using rating sheets for self, peer, or teacher evaluation of paragraphs about results and conclusions are included in Chapter 17, *Evaluating for Success 7: Analyzing and Communicating Data: Descriptive Statistics.* As with a simple report, the components of the rating sheet can be used independently when a specific skill, such as rating a conclusion, is taught. Criteria may also be modified to reflect the school's curriculum or the requirements of a specific science competition.

Related Web Site

http://www.mste.uiuc.edu/stat/stat.html

TABLE 9.5   Writing about Nominal Data

Descriptive information	Concentration of Chemical X			
	0%	10%	20%	30%
Mode	Healthy	Healthy	Unhealthy	Unhealthy
Frequency distribution				
Healthy	10	8	4	2
Unhealthy	0	2	6	8
Number	10	10	10	10

Steps

1. Write a **topic sentence** stating the independent and dependent variables, and a reference to tables or graphs.

2. Write sentences comparing the **measures of central tendency (modes)** of the groups.

3. Write sentences describing the **variation (frequency distribution)** within the groups.

4. Write sentences stating how the data **support the hypothesis**.

Examples

The influence of Chemical X on the health of tomato plants is summarized in Table 9.5.

When plants received 10% or less Chemical X, they remained healthy. At concentrations of 20% X or above, plant health deteriorated.

The greatest variation in plant health was at 20% X, where 4 healthy and 6 unhealthy plants occurred. At concentrations below 20% X, plants were predominantly healthy; at 30% X, the majority of plants were unhealthy.

The data on plant health supported the hypothesis that higher concentrations of Chemical X would adversely affect plant growth.

© 2000 by Kendall/Hunt Publishing Company, Cothron, Giese, & Rezba, *Students and Research*.

TABLE 9.6   Writing about Ordinal Data

Descriptive information	Concentration of Chemical X			
	0%	10%	20%	30%
Median	4	4	2	1
Frequency distribution				
Quality 4	10	6	0	0
Quality 3	0	3	3	0
Quality 2	0	1	7	3
Quality 1	0	0	0	7
Number	10	10	10	10

Steps

1. Write a **topic sentence** stating the independent and dependent variables, and a reference to tables or graphs.

2. Write sentences comparing the **measures of central tendency (medians)** of the groups.

3. Write sentences describing the **variation (frequency distributions)** within the groups.

4. Write sentences stating how the data **support the hypothesis**.

Examples

Leaf quality of plants exposed to various concentrations of Chemical X is summarized in Table 9.6.

High quality leaves, with a rating of 4, were typically found on both control and 10% X plants. At higher concentrations, leaf quality deteriorated, with leaf quality ratings of 2 and 1 on plants grown with 20% X and 30% X.

The greatest variation in leaf color occurred at 10% X where leaf quality ratings of 4, 3, and 2 were assigned. No variation occurred in the control; all leaves received a rating of 4. At 20% X and 30% X, the plants fell into only two categories, 3 and 2, or 2 and 1.

Data on leaf quality supported the hypothesis that Chemical X would impede plant growth.

© 2000 by Kendall/Hunt Publishing Company, Cothron, Giese, & Rezba, *Students and Research*.

TABLE 9.7   Writing a Conclusion

Questions	Examples
1. What was the purpose of the experiment?	The purpose of this experiment was to investigate the effect of various concentrations of Chemical X on the growth of tomato plants.
2. What were the major findings?	At successively higher concentrations of Chemical X, the mean height of the tomato plants decreased and plant health and leaf quality deteriorated. The mean plant height at 10% X (18.1 cm) was greater than the control (15.3 cm), with plants exhibiting similar health. More plants exhibited poor leaf quality in the 10% X group than in the control.
3. Did the data support the hypothesis?	In general, the research data supported the hypothesis that growth of the tomato plants would decrease as the concentration of Chemical X increased.
4. How did your findings compare with other researchers?	Although Crook and Bolton reported that concentrations of 10% X were harmful to radish plants, slightly reduced leaf quality was the only indicator of an adverse effect in this experiment.
5. What possible explanations can you offer for your findings?	Discrepancies in findings could result from different plant species or methods of application. In Crook and Bolton's study, the solution was poured on both plant and soil, whereas in this study the solution was poured only on the soil.
6. What recommendations do you have for further study and for improving the experiment?	Additional studies could be conducted to determine the effect of Chemical X within the 0–20% range on both types of plants and with different methods of application.

© 2000 by Kendall/Hunt Publishing Company, Cothron, Giese, & Rezba, *Students and Research*.

Practice

1. In Chapter 8 you constructed a data table for the Heated Soil Scenario and the Peat Moss Scenario. Write a paragraph of results to accompany each data table.
2. Use the techniques described in this chapter to write a paragraph of results and a conclusion for a laboratory investigation you conducted in class.
3. Use the following data to answer Questions 3A–3F.
    David collected data on the effect of various concentrations of blue metallic pigmentation in paint on the attractiveness of sports cars to teenage buyers.

Very Light Blue Group I	Dark Navy Blue Group II	Light Blue Group III	Blue Group IV	Dark Blue Group V
R	B	R	B	R
R	B	B	R	R
R	B	B	B	R
R	R	B	R	B
B	B	R	R	R
R	B	R	B	B
B	B	R	B	R
R	B	R	R	B
R	R	R	B	R
B = Bought R = Rejected				

A. Which type of data is represented in the data chart?
B. Which level of measurement do the data represent?
C. What is the most appropriate measure of central tendency?
D. What is the most appropriate measure of variation in the data?
E. Construct an appropriate data table.
F. Write an appropriate paragraph of results. (Assume that the researcher hypothesized that the number of sales would increase as the amount of blue pigment increased.)

USING TECHNOLOGY

Integrated Software Programs

In Chapter 8, you conducted an experiment to determine how minerals in water affect suds formation. Use your knowledge of integrated software programs, such as ClarisWorks, Microsoft Office, or WordPerfect, to write a report for "A Sudsy Experience." Expand the simple report, described in Chapter 6, to include summary data tables, graphs, and paragraphs of results.

Specific components of the integrated software package you might use are described below:

Word Processing: general text of document
Drawing a Table: experimental design diagram
Spreadsheet: data tables and graphs/charts
Internet Access: background information for the introduction and conclusion.

© 2000 by Kendall/Hunt Publishing Company, Cothron, Giese, & Rezba, *Students and Research*.

Displaying
Dispersion/Variation
in Data

Objectives

- Make stem-and-leaf plot(s) for a set of data and write a paragraph to summarize findings.
- Make a boxplot for a set of data, construct a data table, and write a paragraph to summarize findings.
- Calculate the variance/standard deviation for a sample and make inferences to a normally distributed population; make a graphical display and data table and write a paragraph to summarize findings.
- Evaluate a set of data and determine the most appropriate measure(s) of dispersion/variation to use.
- Teach students to display the dispersion/variation of data in various ways and to summarize findings through tables, graphs, and paragraphs.

National Standards Connections

- Use appropriate tools and techniques to gather, analyze, and interpret data (NSES).
- Use technology and mathematics to improve investigations and communications (NSES).
- Understand and apply measures of central tendency and variability (NCTM).

In Chapter 8, *Analyzing Experimental Data*, students learned how to summarize experimental data using measures of central tendency and variation. They used simple measures of variation such as range and frequency distribution. In this chapter several techniques are presented to help students expand the possibilities for displaying the dispersion or variation within sets of quanti-

tative data. New techniques include stem-and-leaf plots, boxplots (box and whisker diagrams), and standard deviation/variance. First, have students apply these methods using pencil and paper calculations. Then, if graphing calculators are available, teach them how to calculate statistics and to display the data using these tools. With these additional techniques, students will better understand dispersion/variance in data and will be more able to make informed decisions about differences among groups of experimental data. These new skills will also establish a foundation for students applying the inferential statistics described in Chapter 11.

OVERVIEW OF STATISTICAL AND GRAPHICAL PRESENTATIONS

Before applying these new techniques to experimental data, introduce students to each type with an example. For the example, use the following data on the number of tomatoes produced by 53 plants in a garden.

Sample: 8, 17, 23, 35, 45, 56, 65, 94, 32, 31, 14, 15, 54, 91, 18, 15, 28, 9, 29, 30, 39, 35, 47, 48, 69, 8, 14, 19, 22, 42, 22, 29, 7, 28, 24, 30, 35, 44, 62, 17, 23, 15, 35, 34, 29, 32, 39, 45, 53, 35, 42, 57, 94.

STEM-AND-LEAF PLOTS

Introduce stem-and-leaf plots as a type of bar graph where numeric data are plotted by using the actual numerals in the data to form the graph. They are a quick way to display 25 or more pieces of data. Each piece of data is displayed as two parts: a **stem** and a **leaf.**

In a stem-and-leaf plot, a number, such as 24, is displayed as 2|4; 2 is the stem and 4 is the leaf. Several numbers, such as the tomato data in the 50's (56, 54, 53, 57) would be displayed as 5|6 4 3 7. When ordered, the stem and leaves would appear as 5|3 4 6 7. Include a key, such as 4|8 = 48, as an aid to interpreting the values that the stems and leaves represent.

Data such as the number of tomatoes produced are first put in an **unordered** stem-and-leaf plot (Figure 10.1), which is later converted to an **ordered** stem-and-leaf plot (Figure 10.2). If the data are to be grouped by tens, begin by listing the tens digits in order and drawing a line to the right. These form the **stem** of the graph. Next, go through the tomato data and write the ones digit next to the appropriate tens digit. These are the **leaves.** Stem-and-leaf plots can also be used for larger data. For data in the 100's, for example, the

Stem	Leaf												
0	8	9	8	7									
1	7	4	5	8	5	4	9	7	5				
2	3	8	9	2	2	9	8	4	3	9			
3	5	2	1	0	9	5	0	5	5	4	2	9	5
4	5	7	8	2	4	5	2						
5	6	4	3	7									
6	5	9	2										
7													
8													
9	4	1	4										

Key: 4|8 = 48

Figure 10.1 Unordered Stem-and-Leaf Plot.

Stem	Leaf												
0	7	8	8	9									
1	4	4	5	5	5	7	7	8	9				
2	2	2	3	3	4	8	8	9	9	9			
3	0	0	1	2	2	4	5	5	5	5	5	9	9
4	2	2	4	5	5	7	8						
5	3	4	6	7									
6	2	5	9										
7													
8													
9	1	4	4										

Key: 4|8 = 48

Figure 10.2 Ordered Stem-and-Leaf Plot.

stem is the hundreds digit and the leaves are the tens and ones, such as 463, 479, 482 represented as 4|63, 79, 82. Commas or extra space can be used to separate each leaf.

From the stem-and-leaf plots, answer the following questions about tomato production by various garden plants. Then, check your answers.

1. What values are represented by 4|8, 9|1, 0|8?
2. What was the minimum number of tomatoes produced by a plant? The maximum number? The range in tomato production?
3. What was the most frequent number (mode) of tomatoes produced?
4. Are there **gaps** or empty places in the data?
5. Are there **clusters** or isolated groups of data?
6. What is the general shape of the distribution?

Stem-and-leaf plots are significantly easier for students to construct than bar graphs. In addition, they provide an efficient method of ordering data where individual pieces of data are easily identified. Note in Figures 10.3 a, b, c, and d how effectively stem-and-leaf plots illustrate various data distribution patterns: Bell-shaped (Normal), U-shaped, J-shaped, and Rectangular-shaped.

Answers to Questions on page 116. 1) 48, 91, 8. **2)** Minimum = 7; maximum = 94; range = 87. **3)** Mode = 35. **4)** Gaps in 70's and 80's. **5)** Clusters = 91, 94, 94 **6)** Bell-shaped except for cluster at high end.

a. Bell-shaped (Normal) Distribution: highs balance the lows.

```
1 | 0
2 | 3 3 8
3 | 2 5 6 7
4 | 3 3 4 5 5 8 8 9
5 | 2 5 5 8 9
6 | 4 2 7
7 | 2 3
8 | 1
9 |
```

b. U-shaped (Bi-Model) Distribution: two groups of relatively equal frequency; if studied separately, each group may be bell-shaped.

```
1 | 0 2 2 7 8 9
2 | 3 4 5 6 6
3 | 0 4 5
4 | 2 4
5 | 1 6 7
6 | 2 3 4 5 8
7 | 1 1 2 5 7 8 8
8 |
9 |
```

c. J-shaped Distribution: shows that there is probably a limit to the values of the data.

```
1 | 0 2 2 4 5 7 7 8 9
2 | 1 3 3 6 7 7
3 | 0 4 5 6 8
4 | 3 4 5 3
5 | 2 3 4
6 | 7 8
7 | 2
8 | 2
9 | 0
```

d. Rectangular shaped distribution: uniform distribution of data with values evenly distributed over a range.

```
1 |
2 |
3 | 0 3 3 8 4
4 | 2 3 4 4 7 8
5 | 1 2 4 8 8
6 | 3 4 5 5 9 9
7 | 2 3 4 7 7
8 |
9 |
```

Figure 10.3 Common Data Distribution Patterns.

BOXPLOTS (BOX AND WHISKER DIAGRAMS)

A boxplot is a graphic method of displaying data that is based upon five important components of the data.

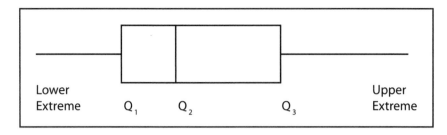

Lower extreme	Minimum value
Lower quartile (Q_1)	Number below which 25% of the values fall
Median (Q_2)	Number which divides the data into half, with 50% of the values falling above the number and 50% below
Upper quartile (Q_3)	Number below which 75% of the numbers fall
Upper Extreme	Maximum value

As shown above, each of these components is displayed by a specific part of the boxplot.

Use this boxplot of tomato production while asking students questions about Figure 10.4.

1. What is the lower extreme? Upper extreme?
2. What is Q_1, Q_2 (median), Q_3?
3. Below which value do 25% of the plants produce? 50% of the plants? 75% of the plants?
4. In what range is the middle 50% of production? Are these middle values for tomato production symmetrically distributed? How can you tell? (Hint: Look at the box.)
5. Where do you have the greatest dispersion or spread in data: the lower quartile or the upper quartile? How can you tell? (Hint: Look at the length of the lines or whiskers.)
6. Because you were estimating from a graph, your numbers may not be exact. How could you use the stem-and-leaf plot to obtain more specific information?

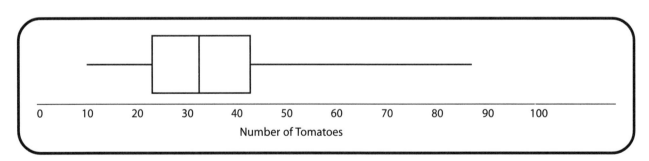

Figure 10.4 Boxplot of Number of Tomatoes Produced.

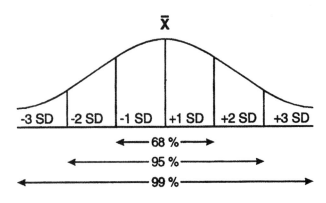

Answers to Question about Figure 10.4: 1) Lower extreme = 7; upper extreme = 94. **2)** Q_1 = 20.5; Q_2 (median) = 32; Q_3 = 45. **3)** 35% are below 20.5; 50% are below 32, and 75% are below 45. **4)** 50% are between 20.5 and 45; values are fairly symmetrical because the two halves of the box are approximately the same lengths. **5)** Greatest dispersion is in the upper quartile where the line (whisker) is much longer. **6)** From the stem-and-leaf plot, you can determine the median. Then, you can determine the median for the lower half (Q_1) and the median for the upper half (Q_3). See if you can do this. How do your answers compare with ours?

Figure 10.5 Normal Distribution of a Population.

STANDARD DEVIATION AND VARIANCE

The standard deviation is a powerful measure of dispersion that is frequently reported in scientific papers. Dispersion is calculated by comparing each individual piece of data in the sample with the mean of the sample. These differences are entered into a specific mathematical formula that is solved to yield the standard deviation. The greater the standard deviation, the more variation or dispersion in the data.

Typically, scientists use the standard deviation when they want to make inferences from a sample to a population. The **sample** is the actual set of objects measured or described; for example, the 53 plants on which the tomatoes were counted. The **population** is the set of "all" tomato plants of the same type grown in similar gardens.

Scientists and statisticians generally assume that populations follow a normal distribution that is illustrated by a bell-shaped curve such as the one shown in Figure 10.5. In a normally distributed population, the mean is found at the peak of the curve. Sixty-eight percent (68%) of the population falls within ± 1 standard deviation (SD) of the mean. Ninety-five percent (95%) of the population falls within ± 2SD of the mean. Similarly, 99% of the population falls within ± 3SD of the mean. Populations frequently consist of very large numbers. For a sample to reflect a population, it must also be sufficiently large. Typically, statisticians recommend that sample sizes be a minimum of 30.

When the standard deviation of a population is graphically presented in this text, the following symbols will be used. In other text, or on graphing calculators, symbols may vary; because a standard system, such as used with boxplots, does not exist (see Figure 10.6).

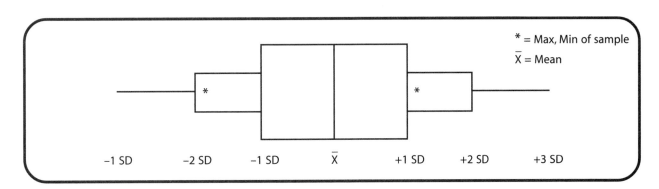

Figure 10.6 Graphic presentation of standard deviations of the entire population.

Below are a data table and graphical presentation of tomato production by the 53 garden plants (see Table 10.1 and Figure 10.7). Use the data table and graph to answer the questions that follow.

1. What is the mean? Standard deviation?
2. What range of tomato production would you expect for 68% of the plants?
3. What range of tomato production would you expect for 95% of the plants? 99% of the plants? Are the values reasonable? Why?
4. Why does the data table and graphical display include values that were not in the sample?

Help students see that the tomato production predicted for the population of "all" tomato plants is not reasonable because there are negative values of –6.1 and –26.9 tomatoes. Remind students that to predict from a sample to a population, a normal distribution is assumed. Refer them back to the stem-and-leaf plot and the boxplot. Although the lower part of the sample appeared "mound shaped," the sample as a whole was not. A gap occurred in the data in the 70's and 80's, with only a few data values occurring in the 90's. The data were **skewed**, or had a tail, at the upper end. Point out this skew in the boxplot where the upper end was substantially longer. When a sample is skewed, stem-and-leaf plots and boxplots are more appropriate ways to display the data. By displaying data in various ways, students can make more informed decisions about the most appropriate way to communicate their findings.

> **Answers about Standard Deviation/Variance of Tomatoes. 1**) Mean = 35.5, SD = 20.8 **2**) 14.7 to 56.3 tomatoes **3**) –6.1 to 77.1 tomatoes; –26.9 to 97.9 tomatoes; values are not reasonable because you cannot have a negative (–) number of tomatoes. **4**) The data table and graph are based on inferences made from the sample to the population and include values that you would expect in the population based upon the dispersion/variation in the sample.

TABLE 10.1 Tomatoes Produced by a Population of Plants

Descriptive information	Number of tomatoes
Mean	35.5
Variance	432.6
Standard deviation	20.8
± 1 SD (68% range)	14.7 to 56.3
± 2 SD (95% range)	–6.1 to 77.1
± 3 SD (99% range)	–26.9 to 97.9
Number	53

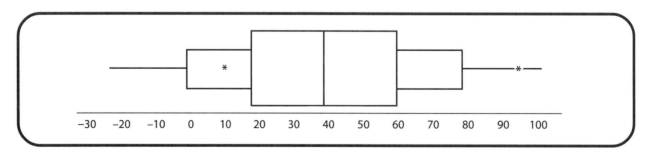

Figure 10.7 Projected Number of Tomatoes Produced by a Population of Plants.

CALCULATING DISPERSION/ VARIATION IN EXPERIMENTAL DATA

Introduce the investigation by showing students straws with paper wrappers. Ask: "Have you ever played with the paper, maybe pushing it tightly together at one end?" If so, you know that you get a "worm-like" piece of paper. Use the proce-dure in Investigation 10.1, *Paper Worms*, to turn these "worms" into an experiment.

Below are data on paper worms collected by a class of 30 students (see Table 10.2). The data will be used to calculate various measures of dis-persion/variation. Follow these procedures to help students analyze their data from the paper worm experiment.

TABLE 10.2   Length of Paper Worms (mm)

0 drops of water (dry)						4 drops of water (wet)					
30	35	74	49	55	20	65	55	114	89	80	55
38	40	50	65	32	40	85	40	91	105	75	67
24	38	44	40	58	66	60	65	81	90	65	97
40	64	37	34	38	44	57	95	87	90	55	76
48	70	43	49	37	45	85	110	75	70	37	74

INVESTIGATION 10.1 • Paper Worms

Question

How do paper worms respond to water?

Materials for Each Student

- Straw with paper wrapper
- Small container of water
- Metric ruler
- Eyedropper or straw

Safety

- Dispose of straws after experimenting.
- Do not place straws in mouth.

Procedure

1. Hold a straw vertically. Tear the paper at the top end of the straw. Carefully, push the paper tightly together down to the bottom end. Try not to tear the paper. When you have finished, remove the paper worm.

2. Measure the length of the paper worm in millimeters (mm). If the paper worm is curved, measure the shortest distance between the two ends. Record the data.

3. Using an eyedropper (or your straw), add 4 drops of water along the length of the worm. Measure the length of the worm (mm). Record your data.

	0 drops water (dry)	4 drops water (wet)
Length (mm)	_____	_____

4. Enter the data in a class data table. If necessary, repeat the experiment so that your class has 30 or more trials.

Analyzing the Data

1. Make stem-and-leaf plots of both the wet and dry paper worm data. Write a paragraph describing the data.

2. Make boxplots of both the data sets. Display the information in a data table and a graph. Write a paragraph about the results.

3. Calculate the standard deviation/variance for a population of wet and dry paper worms. Display the information in both a data table and a graph. Write a paragraph about the results.

4. What are appropriate ways to display your experimental findings? Why?

5. Write a report about your experiment with paper worms.

(continued on the following page)

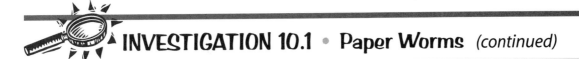

INVESTIGATION 10.1 • Paper Worms *(continued)*

Extending Your Learning

1. What type of energy did you use to make the paper worm?
   What type of energy did the paper worm have?
   What type of energy did the water possess?
   What energy transformations occurred as you added water to the paper worm?

2. What is the chemical composition of paper? How did the water interact with the chemicals in the paper?

3. How is tissue paper, the type of paper around the straw, manufactured? Do you think you would get the same results with other types of paper? Why?

4. What other factors affect the action of paper worms? Design and conduct experiments to test your hypotheses.

USING TECHNOLOGY ·

1. In the **STAT** mode of your calculator, enter the dry worm and wet worm measurements in Lists 1 and 2. Sort the measurements in ascending order. Use the ordered data to make ordered stem-and-leaf plots. (See Appendix A, *Using Technology*, for additional help in using the graphing calculator.)

2. In the **STAT** mode select CALC (for calculate) and then 1 VAR (for 1-variable statistics). Depending on the brand of your calculator, you will need to enter the desired list number (e.g., L1), or SET the 1-variable x-list to the desired list number before selecting 1 VAR. *Repeat* the selection process for the wet worm data by changing the list number to List 2. Obtain 1-variable statistics for each list of data including mean, median, quartiles, minimum, maximum, standard deviation, and number. Use this information to make data tables and graphical displays.

3. Set up a different graph for the dry worm and wet worm data sets. In setting up the first graph, select boxplot as your graph type and List 1 for the x-list and 1 for the frequency. Repeat to set up the second graph, but enter List 2 for the x-list. Graph both boxplots simultaneously.

4. Press **Trace** and use the left and right arrow keys to highlight such values as the minimum, 1st quartile, median, 3rd quartile, and maximum. To make comparisons between the two boxplots, use the up or down arrows to toggle between graphs.

5. Link your graphing calculator to a computer and download these boxplots to save, enlarge, or print them.

STEM-AND-LEAF PLOTS

In the beginning of this chapter, a stem-and-leaf plot was made for one set of data. In this experiment, there are two sets of data, one for 0 drops of water and one for 4 drops of water. When two sets of data exist, a back-to-back stem-and-leaf plot can be constructed. If more than two sets of data exist, you can construct multiple stem-and-leaf plots. To facilitate data interpretation, use the same scale for the "stems" and place the plots beside or underneath each other.

Step 1:   Determine the minimum and maximum values of the data. For 0 drops of water, the values are 20 mm and 74 mm. For 4 drops of water, the values are 37 mm and 114 mm. In constructing the back-to-back stem-and-leaf plot, use the minimum and maximum values from both data sets, e.g. 20 mm and 114 mm.

Step 2:   In the center of the page, write the "stems" vertically. Draw a dark line on both sides of the stem.

Step 3:   On the left side of the stem, write the "ones place" values of the data for 0 water drops. On the right side of the stem, write the "ones place" values of the data for 4 water drops. This is an unordered stem-and-leaf plot (see Figure 10.8).

Step 4:   Make an ordered stem-and-leaf plot by arranging the leaves in order from the "inside" to the "outside." Be sure to include a key (see Figure 10.9)

Step 5:   Write a paragraph about the data. In the paragraph provide the following information:

- Minimum and maximum values and range
- Most frequently occurring value (mode)
- Shape-of-the-data
- Existence of clusters and gaps
- Support for hypothesis.

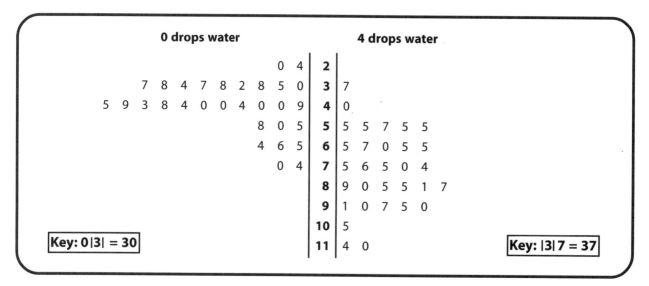

Figure 10.8 Back-to-Back Unordered Stem-and-Leaf Plots for Impact of Water on the Length of Paper Worms.

04-10-2007 12:52PM
Item(s) checked out to Hood, Trevor.

TITLE: Science probe 10 / Author team, J
BARCODE: 30020004166726
DUE DATE: 25-10-07

TITLE: Minds on science : the art of tea
BARCODE: 30020001483108
DUE DATE: 25-10-07

TITLE: Ready-to-use science proficiency
BARCODE: 30020004163947
DUE DATE: 25-10-07

TITLE: Invitations to science inquiry /
BARCODE: 30020004169746
DUE DATE: 25-10-07

TITLE: Students and research : practical
BARCODE: 30020004031839
DUE DATE: 25-10-07

University of Northern BC
Weller Library

Stem-and-Leaf Plot

	...ater							4 drops water					
			4	0	**2**						
	7	5	4	2	0	**3**	7				
	3	0	0	0	0	**4**	0				
			8	5	0	**5**	5	5	5	7	
			6	5	4	**6**	0	5	5	5	7
				4	0	**7**	0	4	5	5	6
						8	0	1	5	5	7	9
						9	0	0	1	5	7
						10	5				
						11	0	4			

Key: |3|7 = 37

Paragraph

...rom 20 mm to 74 mm, with the most typical length being 40 mm.
...ed from 37 mm to 114 mm, with the most typical measurements
...e data formed a very steep mound; most of the measurements were
...s a much flatter mound, with the center part being rectangular. The
...d increase worm length. Typical lengths shifted from the 20's and

...af Plots and Paragraph for Impact of Water on Length of Paper

BOXPLOTS (BOX AND WHISKER DIAGRAMS)

Remind students of their experience with a box-plot of tomato production by various garden plants. Help them construct boxplots using the ordered stem-and-leaf plots for the dry and wet paper worms and the procedures described on page 126 (see Figure 10.10).

0 drops water										**4 drops water**				
						4	0	**2**					
	8	8	8	7	7	5	4	2	0	**3**	7			
9	9	8	5	4	4	3	0	0	0	0	**4**	0		
						8	5	0	**5**	5	5	5	7	
						6	5	4	**6**	0	5	5	5	7
							4	0	**7**	0	4	5	5	6
									8	0	1	5	5	7	9
									9	0	0	1	5	7
									10	5			
									11	0	4		

Key: 0|3| = 30   Key: |3|7 = 37

Figure 10.10 Ordered Stem-and-Leaf Plots for Dry and Wet Paper Worms.

Step 1:   Find the median or mid-point of the data. Both sets of data contain 30 values. The median will be the point halfway between the 15th and 16th value (see underlined values).

Median (Q_2) for 0 drops $\dfrac{40 \text{ mm} + 43 \text{ mm}}{2} = 41.5 \text{ mm}$

Median (Q_2) for 4 drops $\dfrac{75 \text{ mm} + 76 \text{ mm}}{2} = 75.5 \text{ mm}$

Step 2:   Find the median of the lower half of the data (Q_1). Because 15 values are found in the lower half, Q_1 is the 8th number (see shaded values).
Lower quartile (Q_1) for 0 drops   37 mm
Lower quartile (Q_1) for 4 drops   65 mm

Step 3:   Find the median of the upper half of the data (Q_3). Fifteen values are found in the upper half. The median of these values is the 8th number because 7 numbers fall above and below it (see shaded valued).
Upper quartile (Q_3) for 0 drops   50 mm
Upper quartile (Q_3) for 4 drops   90 mm

Step 4:   Find the extreme values of the data, that is the minimum and the maximum values.
0 drops   Min = 20 mm Max = 74 mm
4 drops   Min = 37 mm Max = 114 mm

Step 5:   Develop a scale for plotting the values. Use the techniques previously described in Chapter 5.

$$\frac{\text{Max} - \text{Min}}{5} = \frac{114 \text{ mm} - 20 \text{ mm}}{5} = \frac{94 \text{ mm}}{5} = 18.8 \text{ mm} \sim 20 \text{ mm}$$

Generally, mathematicians graph boxplots on a horizontal number line. Graphing calculators also plot horizontally. Often scientists will plot vertically, so that the independent variable (x) and the dependent variable (y) are on the expected axes. Either method is appropriate as long as the variables are clearly labeled.

Step 6:   Plot points for the upper quartile (Q_3) and the lower quartile (Q_1). Make a box. Draw a vertical line across the box at the median (Q_2). Draw a line or "whisker" from each quartile to the extreme values. (See Table 10.2.)

Step 7:   Find the **interquartile ranges (IQR).** This is the difference between Q_3 and Q_1.
0 drops   50 mm – 37 mm = 13 mm
4 drops   90 mm – 65 mm = 25 mm

Step 8:   Determine if **outliers**—very small or very large points—exist. Such points are greater or less than 1.5 times the interquartile range.
0 drops   IQR = 13 mm   13 mm × 1.5 = 19.5 mm
             Lowest reasonable value = Q_1 – 19.5 mm =
                      37 mm – 19.5 mm = 17.5 mm

             Highest reasonable value = Q_3 + 19.5 mm =
                      50 mm + 19.5mm = 69.5 mm

Outliers exist at the upper end: 70 and 74 mm. No outliers exist at the lower end. Place an asterisk (*) on the whisker where these points exist.

4 drops   IQR = 25 mm   25 mm $\times$ 1.5 = 37.5 mm

                Lowest reasonable value = Q_1 – 37.5 mm =
                                         65 mm – 37.5 mm = 27.5 mm

                Highest reasonable value = Q_3 + 37.5 mm =
                                         90 mm + 37.5 mm = 127.5 mm

No outliers exist.

Step 9:   Write a paragraph about the boxplots that includes the following information (see Table 10.3 *Summary Data Table and Boxplots for the Impact of Water on the Length of Paper Worms (mm)*):

- Introductory sentence
- Comparison of medians
- Comparison of boxes and interquartile range (middle 50% of values)
- Symmetry of distribution
- Range of values and outliers
- Support for hypothesis.

TABLE 10.3   Summary Data Table, Boxplots, and Paragraph for the Impact of Water on the Length of Paper Worms (mm)

Descriptive information	Number of water drops	
	0	4
Median (Q_2)	41.5	75.5
Dispersion		
Minimum	20	37
Q_1	37	65
Q_3	50	90
Maximum	74	114
Interquartile range	13	25
Reasonable lowest value	17.5	27.5
Reasonable highest value	69.5	127.5
Outliers	70, 74	None
Number	30	30

Boxplots

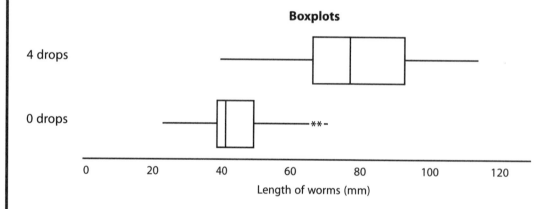

Paragraph

The effect of water on the length of paper worms is displayed in the table above and accompanying boxplots. When water was added, the median length of the paper worms increased from 41.5 mm to 75.5 mm. For both dry and wet worms, the middle 50% of lengths were fairly symmetrical. Almost twice as much dispersion occurred in wet worms as evidenced by the interquartile range: 25 mm versus 13 mm. Overall, wet worms also showed greater dispersion, a total range of 77 mm versus 54 mm. For both sets of worms, dispersion was greatest in the upper quartile. Two outliers, 70 mm and 74 mm, occurred in the dry worm data. The data supported the hypothesis that water would increase worm length, because the medians were approximately two quartiles apart, with the upper 50% of dry worm lengths overlapping the lower 50% of wet worm lengths.

STANDARD DEVIATION AND VARIANCE

Previously you looked at the standard deviation of tomato production by various garden plants. These values were calculated by comparing each individual value with the mean of the sample. Four basic steps are involved in calculating the variance and standard deviation of a sample, as shown in Table 10.4 *Calculating the Variance and Standard Deviation of the Dry Worm Lengths (mm)* for the dry worm data, and summarized below.

Step 1: Find the difference between each individual value (X_i) and the mean ($\overline{X}$).

	Examples	Value 1	20 mm – 44.9 mm = –24.9 mm
		Value 19	45 mm – 44.9 mm = 0.1 mm
		Value 29	70 mm – 44.9 mm = 25.1 mm

Step 2: Square the difference between each individual value and the mean.

	Examples	Value 1	$(-24.9 \text{ mm})^2 = 620.01 \text{ mm}^2$
		Value 19	$(0.1 \text{ mm})^2 = .01 \text{ mm}^2$
		Value 29	$(25.1 \text{ mm})^2 = 630.01 \text{ mm}^2$

Step 3: Find the sum of the squared differences $(X_i - \overline{X})^2$. Divide this difference by the degrees of freedom; that is, the number in the sample (n) minus 1. The dividend is the unbiased estimate of the variance of the population.

$$\text{Variance} = \frac{620.01 \text{ mm}^2 + 436.81 \text{ mm}^2 \ldots\ldots + 846.81 \text{ mm}^2}{30 - 1}$$

$$= \frac{5464.7 \text{ mm}^2}{29}$$

$$= 174.6 \text{ mm}^2$$

Step 4: Find the square root of the variance. This number is the unbiased estimate of the standard deviation of the population.

$$\text{Standard deviation} = \sqrt{174.6 \text{ mm}^2} = 13.2 \text{ mm}$$

Step 5: Prepare an appropriate data table, graphic display, and paragraph (see Table 10.5).

Statisticians use a biased and unbiased measure of the standard deviation of the population. A **biased** measure is found by dividing by the number (n) in the sample; it is generally too low an estimate of the population estimate. The **unbiased estimate** is found by dividing by the degrees of freedom (n – 1). In this text, the unbiased estimate of variance and standard deviation will be used. On calculators, this generally has one of the following symbols: Texas Instruments (Sx or σ n − 1) and Casio (x σ n − 1).

TABLE 10.4   Calculating the Variance and Standard Deviation of the Dry Worm Lengths (mm)

Trial	Individual Values (X_i)	Mean ($\bar{X}$)	Step 1 Individual Value – Mean ($X_i - \bar{X}$)	Step 2 Squared difference ($X_i - \bar{X}$)^2
1	20	44.9	−24.9	620.01
2	24	44.9	−20.9	436.81
3	30	44.9	−14.9	222.01
4	32	44.9	−12.9	166.41
5	34	44.9	−10.9	118.81
6	35	44.9	−9.9	98.01
7	37	44.9	−7.9	62.41
8	37	44.9	−7.9	62.41
9	38	44.9	−6.9	47.61
10	38	44.9	−6.9	47.61
11	38	44.9	−6.9	47.61
12	40	44.9	−4.9	24.01
13	40	44.9	−4.9	24.01
14	40	44.9	−4.9	24.01
15	40	44.9	−4.9	24.01
16	43	44.9	−1.9	3.61
17	44	44.9	−0.9	0.81
18	44	44.9	−0.9	0.81
19	45	44.9	0.1	0.01
20	48	44.9	3.1	9.61
21	49	44.9	4.1	16.81
22	49	44.9	4.1	16.81
23	50	44.9	5.1	26.01
24	55	44.9	10.1	102.01
25	58	44.9	13.1	171.61
26	64	44.9	15.1	364.81
27	65	44.9	20.1	404.01
28	66	44.9	21.1	445.21
29	70	44.9	25.1	630.01
30	74	44.9	29.1	846.81

$$\text{Mean} = \frac{1347 \text{ mm}}{30} = 44.9 \text{ mm}$$

Step 3 $\text{Variance} = \frac{5464.7 \text{ mm}^2}{30 - 1} = 174.6 \text{ mm}^2$

Step 4 $\text{SD} = \sqrt{174.6 \text{ mm}^2} = 13.2 \text{ mm}$

TABLE 10.5   The Effect of Water on the Standard Deviation/Variance of Paper Worm Lengths (mm)

Data Table

Descriptive information	Number of water drops	
	0	**4**
Mean	41.5	76.3
Standard deviation (unbiased)	13.2	19.1
± 1 SD (68% range)	31.7 to 58.1	57.2 to 94.5
± 2 SD (95% range)	18.5 to 71.3	38.1 to 114.5
± 3 SD (99% range)	5.3 to 84.5	19.0 to 133.6
Number	30	30

Graphic Display

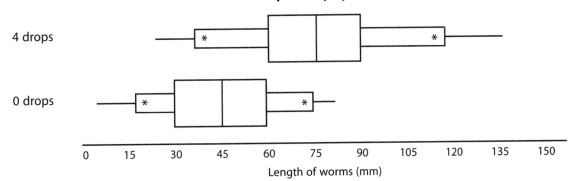

Paragraph

The impact of water on the length of paper worms is displayed in the above data table and graph. After water was added, the mean length of the worms increased from 44.9 mm to 76.3 mm. Greater dispersion occurred in the wet paper worms. Projections for the populations of "all" worms showed that the means were approximately 2 SD apart. The data supported the hypothesis that water would increase the length of the worms.

MAKING DECISIONS ABOUT DISPERSION/VARIATION IN DATA

In Chapter 8, students used a table to make decisions about appropriate measures of central tendency and dispersion. They began by classifying the independent variable as continuous or discrete. Then, they classified the dependent variable as quantitative (continuous/discrete), qualitative (ordinal), or qualitative (nominal). Table 10.6 *Making Decisions* *about Measures of Central Tendency and Dispersion/Variation* is an expanded version of the table that includes the new measures of dispersion/variation and graphical displays that students learned about in this chapter.

Have students use the information in Table 10.6 to determine appropriate statistics for Investigation 10.2, *Watery Statistics*. Then, have them complete the investigation and write a report. Students will also find the table helpful in solving the practice problems at the end of this chapter.

TABLE 10.6   Making Decisions about Measures of Central Tendency and Dispersion/Variation

Levels of Independent Variable	Dependent Variable	Central Tendency	Dispersion/ Variation	Graphic Displays
Continuous	Quantitative (continuous/ discrete)	Mode Median Mean	Range Variance Standard Deviation	Bar Graph Line Graph, Line-of-Best Fit (continuous) Line Graph, Broken Line (discrete) Stem-and-Leaf Plots Boxplots Standard Deviation
	Qualitative (ordinal)	Mode Median	Frequency distribution	Bar Graph (medians) Frequency distribution
	Qualitative (nominal)	Mode	Frequency distribution	Frequency distribution
Discrete	Quantitative (continuous/ discrete)	Mode Median Mean	Range Variance Standard Deviation	Bar Graph Stem-and-Leaf Plots Boxplots Standard Deviation
	Qualitative (ordinal)	Mode Median	Frequency distribution	Bar graph (medians) Frequency distribution
	Qualitative (nominal)	Mode	Frequency distribution	Frequency distribution

INVESTIGATION 10.2 • Watery Statistics

Question

How does surface affect the size of water drops?

Materials for Each Student

- Small container of water
- Dropping device such as eyedropper, baster, squeeze bottle, straw
- Metric ruler
- Various surfaces: wax paper, aluminum foil, plastic wrap

Procedures

1. Fill the dropper with water.
2. Hold the dropper 5 cm above a sheet of wax paper, with the shiny side up; make a drop on the surface.
3. Measure the diameter or longest dimension of the drop to the nearest mm. Record the data.
4. Repeat Steps 2–3 with aluminum foil (shiny side) and plastic wrap (top/shiny side).

Diameter of drops (mm)

Wax paper _____   **Al foil** _____   **Plastic wrap** _____

5. Record your data in a class data table. Be sure that at least 30 sets of data are entered. If there are fewer than 30 trials, then some students should repeat Steps 1–4 to obtain sufficient trials.

Analyzing Your Data

1. Make stem-and-leaf plots of the data and write a paragraph to summarize findings.
2. Make boxplots. Summarize findings in a data table and paragraph.
3. Calculate the standard deviation/variance. Make an appropriate data table and write a paragraph about results.
4. Write a report about the experiment.

(continued on the following page)

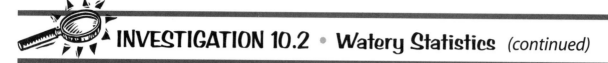

INVESTIGATION 10.2 • Watery Statistics *(continued)*

USING TECHNOLOGY ·

1. In the **STAT** mode of your calculator, enter the water drop measurements in Lists 1, 2, and 3. Sort the measurements in ascending order. Use the ordered data to make ordered stem-and-leaf plots. (See Appendix A, *Using Technology*, for additional help in using the graphing calculator).

2. In the **STAT** mode select CALC (for calculate) and then 1 VAR (for 1-variable statistics). Depending on the brand of your calculator, you will need to enter the desired list number (e.g., L1), or SET the 1-variable x-list to the desired list number before selecting 1 VAR. *Repeat* the selection process for each set of data by changing the list number. Obtain 1-variable statistics for each list of data including mean, median, quartiles, minimum, maximum, standard deviation, and number. Use this information to make data tables and graphic displays.

3. Set up a different graph for each set of data you want displayed. In setting up the first graph, select boxplot as your graph type and List 1 for the x-list and 1 for the frequency. Repeat to set up the second and third graphs, but enter List 2 or List 3 for the x-list. Graph all three simultaneously.

4. Press **Trace** and use the left and right arrow keys to highlight such values as the minimum, 1st quartile, median, 3rd quartile, and maximum. To make comparisons among the three different boxplots, use the up or down arrows to toggle between graphs.

5. Link your graphing calculator to a computer and download these boxplots to save, enlarge, or print them.

Extending Your Learning

1. What is adhesion? Cohesion? Capillary action?

2. What is the chemical composition of wax paper, aluminum foil, and plastic wrap? Use this information to explain differences among the water drops.

3. What other variables could you measure or describe? How could you operationally define them? What type of measures of central tendency and variation/dispersion would be appropriate?

4. What other factors affect the behavior of various liquids on various surfaces? Design experiments to test your hypotheses.

© 2000 by Kendall/Hunt Publishing Company, Cothron, Giese, & Rezba, *Students and Research*.

EVALUATING AND REINFORCING SKILLS

Evaluation suggestions for paper-pencil tests are included in Chapter 15. Items require students to construct data tables, graphic displays, and paragraphs about stem-and-leaf plots, boxplots, and standard deviation/variance. In Chapter 16, the various components of Table 16.8 *Analyzing and Communicating Measures of Dispersion/Variation* can be used to assess student products.

Related Web Sites

www.statcan.ca
http://www.mste.uiuc.edu/stat/stat.html
www.lib.umich.edu/libhome/
    Documents.center/stats.html
www.bls.gov/oreother.htm
http://www.scri.fsu.edu/
    ~dennisl/CMS/sf/sf_
    details.html
    (standard deviation)

REFERENCES

Glenberg, A. (1996). *Learning from data: An introduction to statistical reasoning* (2nd ed.). Mahwah, NJ: Lawrence Erlbaum Associates.

Landwehr, J.M. & Watkins, A.E. (1994). *Exploring data.* A component of the Quantitative Literacy Series. Palo Alto, CA: Dale Seymour Publications.

Landwehr, J.M., Swift, J., & Watkins, A.E. (1998). *Exploring surveys and information from samples.* A component of the Data-Driven Mathematics Series. Palo Alto, CA: Dale Seymour Publications.

McClave, J.T., Dietrich, Frank, H. II, & Sincich, T. (1997). *Statistics* (7th ed.). Upper Saddle River, NJ: Prentice-Hall, Inc.

Shavelson, R.J. (1996). *Statistical reasoning for the behavioral sciences* (3rd ed.). Boston: Allyn & Bacon.

Yates, D.S., Moore, O.S., McCabe, G.P. (1999). *The Practice of Statistics: TI-83 Graphing Calculator Enhanced.* New York: W.H. Freeman and Company.

Practice

For the following data set: a) determine appropriate measures of central tendency and dispersion/variation, b) construct stem-and-leaf plots and write a paragraph about findings, c) construct boxplots, data table, and paragraph of results, d) calculate the standard deviation/variance and make appropriate data tables, graphical displays, and paragraphs of results, and e) select the most appropriate method for displaying dispersion/variation and justify your answer.

1. Data on the Effect of Various Percentages of Nitrogen in Fertilizer on the Height of Field Corn Plants (m)

	Percentage (%) of Nitrogen in Fertilizer											
Trials	Control (0%)			5%			10%			15%		
1–3	1.9	0.9	0.8	2.2	2.3	2.4	3.3	2.7	2.5	3.9	3.2	3.1
4–6	1.5	1.5	2.0	2.1	2.6	2.7	3.1	2.8	2.1	3.2	3.0	3.6
7–9	1.0	2.0	1.4	2.9	2.5	2.3	3.0	2.9	2.4	3.7	3.5	3.8
10–12	1.4	0.9	1.3	2.6	2.8	2.6	3.3	3.2	3.3	3.5	3.2	3.2
13–15	0.9	1.4	1.3	2.2	2.6	2.7	3.0	2.9	2.6	3.6	3.9	3.9
16–18	0.8	1.2	1.8	2.5	3.0	2.8	3.0	3.4	2.5	3.7	3.0	3.1
19–21	1.2	1.3	1.3	2.6	2.3	2.6	3.0	3.4	3.1	3.4	4.1	3.5
22–24	0.5	1.4	1.9	2.6	2.8	2.4	2.7	3.2	3.9	3.2	3.4	3.2
25–27	1.5	0.9	1.5	3.1	2.5	2.7	2.7	2.4	3.4	3.8	3.6	3.7
28–30	1.3	1.8	1.6	2.8	2.4	2.5	2.8	3.3	3.5	3.1	3.7	3.5
31–33	1.7	0.8	1.3	2.8	2.9	2.7	3.3	2.7	3.2	3.4	3.9	3.4
34–35	1.0	1.7		3.6	2.8		2.9	3.0		3.5	3.7	

2. Ms. Goldfarb's students thought that sometimes they didn't get their money's worth at a local fast food restaurant when they bought a jumbo order of french fries. Each server seemed to vary a lot in the number of french fries they put in each order. The students designed a study to answer the following question: Does the number of french fries in

(continued on the following page)

Practice (continued)

the orders served by an individual server vary? On a particular day they asked people to count the number of french fries they were served. The data the students collected were:

Server 1							
32	38	33	32	35	32	34	31
34	34	37	33	36	38	34	37
36	35	36	36	31	33	35	
35	31	34	34	33	32	37	

Server 2								
29	31	34	31	31	32	31	32	29
34	33	30	28	30	29	33	34	32
28	29	28	32	34	33	28	28	30
33	32	30	29	30	33	34	31	

Server 3							
32	32	29	30	33	33	33	33
32	33	30	33	33	31	29	32
33	32	33	32	32	33	33	
31	30	31	33	31	32	33	

Server 4								
34	29	32	33	36	29	30	30	
31	34	30	35	32	31	28	29	
33	30	34	34	30	35	31		
36	35	30	28	34	34	33		

A. Use a leaf-stem plot for each set of the servers' data to determine if any of the sets of data for a server form a J distribution.

B. Use a leaf-stem plot for each set of the servers' data to determine if any of the sets of data for a server form a normal distribution.

C. Use a leaf-stem plot for each set of the servers' data to determine if any of the sets of data for a server form a bimodal or U distribution.

D. Use a leaf-stem plot for each set of the servers' data to determine if any of the sets of data for a server form a rectangular distribution.

E. What is the range, mean, median, Q_1, Q_3, and standard deviation for each server?

	Server 1	Server 2	Server 3	Server 4
range				
mean				
median (Q_2)				
Q_1				
Q_3				
SD				

F. Which server had the greatest variation in the number of french fries served? How can you tell?

G. Which server had the least variation in the number of french fries served? How can you tell?

© 1998 PhotoDisc, Inc.

CHAPTER 11

Determining Statistical Significance

Objectives

- Determine the level of significance and degrees of freedom for a statistical test and use these concepts to explain the probability of error.
- Identify the appropriate inferential statistic to use for a given set of data.
- Construct an appropriate data table and graph to communicate data.
- Teach students the fundamentals of inferential statistics.
- Teach students to calculate and interpret a *t* test and a chi-square test.
- Teach students to write a paragraph to communicate the results of a statistical test.

National Standards Connections

- Formulate and revise scientific explanations and models using loci and evidence (NSES).
- Understand sampling and recognize its role in statistical claims (NCTM).
- Test hypotheses using appropriate statistics (NCTM).

By the end of February, Mrs. Smith's biology students completed their experiments and began to organize data into appropriate tables and graphs. Unanswered questions remained. Is the research hypothesis supported by the data? Did the experimental treatment make a difference? What recommendations will emerge?

Amy rapidly completed her qualitative data table comparing the height of 75 ornamental mimosa seedlings that were forced to close twice daily with a control group consisting of an equal number of plants. For three months, the experi-

mental seedlings, age one year, had been stroked at 7:00 A.M. and 7:00 P.M. to force closure. Heights of the experimental and control plants were measured and recorded. Amy knew that the thigmotrophic response of plants provided protection and long-term survival; however, she hypothesized that the short-term effects upon growth would be negative because of increased energy expenditures. Amy reported that stroked plants exhibited a mean height of 80 cm, as compared with 87 cm for the control. Slightly greater variation in plant height occurred within the stroked group (standard deviation = 8.2 cm) than within the control (standard deviation = 5.8 cm). Amy concluded that the data supported the research hypothesis and recommended that the thigmotrophic response of houseplants, not typically subject to adverse conditions, be eliminated through selective breeding.

After Ben learned that acetylene, a hydrocarbon gas, hastened the ripening of fruit, he hypothesized that other hydrocarbons produced by the incomplete combustion of fossil fuels would have a similar effect as shown in Table 11.1. Ben methodically counted the number of apples displaying various colors when ripened in a smoke-filled atmosphere as compared with a normal (control) atmosphere. Although apples ripened in normal and smoke-filled atmospheres had the same mode, light red, substantial variation occurred in fruit color.

TABLE 11.1   Color of Apples in Various Atmospheres

	Color of Apples				
	Green	Pink	Light Red	Medium Red	Dark Red
Smoke-filled atmosphere (Experimental group)	5	15	35	25	20
Normal atmosphere (Control group)	10	30	45	10	5

Because more ripened fruit occurred in the smoke-filled atmosphere, Ben concluded that the research hypothesis was supported and that fossil fuel pollution could affect patterns of fruit ripening.

Based upon descriptive statistics, both Amy and Ben concluded that their research hypotheses were supported and cited applications for other groups of mimosa plants and apple trees. However, were the findings reported by Amy and Ben **statistically significant?** What is the probability that the differences occurred by chance and were not a result of the experimental treatment? What inferences can be made from the samples of mimosa plants and apples that could apply to populations of these items? To answer these questions, inferential, rather than descriptive statistics must be used.

Descriptive statistics consist of mathematical procedures that report important characteristics of data, including measures of central tendency and variation. These procedures are described in Chapter 8. **Inferential statistics** expand the researcher's framework from a small group, the sample, to the entire group, the population. Various mathematical procedures, statistical tests, exist for determining the probability that observed differences result from the experimental treatment, rather than from chance. Just as with descriptive statistics, different inferential statistical tests are used with quantitative and qualitative data. Two frequently used tests are the *t* test for quantitative data and the **chi-square** test for qualitative data. The *t* test is used to determine whether significant differences exist between means; whereas, the chi-square test evaluates the significance of differences between frequency distributions. Table 11.2 *Appropriate Statistics for Various Data* summarizes the major descriptive and inferential statistics used with quantitative and qualitative data. This chapter discusses basic principles of inferential statistics, calculations of the *t* test and chi-square test, and reporting of statistical findings.

POPULATIONS AND SAMPLES

A scientist uses samples as a vehicle for investigating a population. The sample, the population from which the sample was drawn, and the target population represent successively larger units of study that are operationally defined by the researcher:

- **Sample:** the specific portion of the population that is selected for study, for example, *the 150 mimosa seedlings used in Amy's study;*
- **Sampled Population:** the population from which the sample was drawn, e.g., *all the mimosa seedlings in the nursery from which Amy obtained her mimosa seedlings;*

TABLE 11.2 Appropriate Statistics for Various Data

Category	Analysis of data	Quantitative data (Continuous)	Qualitative data (Discrete-Categorical)
Descriptive Statistics	Measure of central tendency	Mean	Median Mode
	Measure of variation	Range Variance Standard deviation	Frequency distribution
Inferential Statistics	Statistical test	*t* Test	Chi-Square

- **Target Population:** all units (persons, things, experimental outcomes) of the specific group whose characteristics are being studied, such as *all the mimosa seedlings of the same species.*

The validity of an experiment depends on a precise definition of the population and careful sampling of the defined population. Populations may be very large, such as all white oak trees within the southeastern United States, or very small, such as all students in a given chemistry class. Populations may consist of complete entities (trees, people, motors) or subdivisions within these entities (xylem cells, hearts, or fan belts). For validity, the sampled population must be similar to the target population. If the mimosa seedlings in the nursery Amy used differed greatly from other mimosa seedlings of the same species, the findings would be invalid. Selection of the sample is critical. Random samples in which every individual member of the population has an equal chance of being included are preferred and are an underlying assumption of many statistical tests. Tables of random numbers, generated by computers or located in standard reference texts, may be used in drawing samples. To draw a random sample from 2,000 mimosa seedlings in a nursery, Amy would assign a four digit number, starting with 0001, to each plant and then consult a table of random numbers to determine the specific plants to include in the 150-plant sample.

Bias can occur when samples are drawn so that all numbers do not have an equal chance for inclusion. For example, bias would exist if Amy selected all mimosa plants for the experimental group from one part of the greenhouse and plants for the control group from another section. Variations between growth conditions in the two parts of the greenhouse or among the genetic stock giving rise to the plants could introduce such bias.

Confidence in experimental findings increases with additional repeated trials. Likewise, confidence in inferences from the sample to the population increases with larger random samples. In a large random sample, a continuous variable such as plant height tends to fall into a bell-shaped frequency distribution, or normal curve, with the sample mean ($\bar{X}$) approximating the population mean (μ). Multiple random samples may be drawn from the same population. With descriptive statistics, the mean, variance, and standard deviation of the sample means can be calculated. The means ($\bar{X}$) of the samples tend to fall into a normal curve with the mean of the sample means ($\bar{\bar{X}}$) approximating the population mean (μ). Similarly, the standard deviation of the samples approximates the standard deviation (σ) of the population. The special term, **standard error,** is used for the standard deviation of the sample means. Relationships among a sample, multiple samples, and the population are depicted in Table 11.3 *Relationships Among Samples of Populations.*

HYPOTHESES AND SIGNIFICANCE

Scientists design experiments to determine if a specific research hypothesis or set of hypotheses is supported by the data. A **research hypothesis** is derived from the literature review and suggests

TABLE 11.3   Relationships Among Samples of Populations

Sample (Mimosa plants)	Multiple Samples (100 mimosa plants/sample)	Population (All targeted mimosa plants)

Increased Confidence →

10 plants

100 plants

1,000 plants

Sample mean $= \bar{X} = \dfrac{\Sigma X_i}{n}$

Sample variance $= s^2 = \dfrac{\Sigma(X_i - \bar{X})^2}{n - 1}$

Sample standard deviation $= s = \sqrt{s^2}$

Frequency distribution of plant heights in sample

Increased Confidence →

① ② ③ . . . . ⑩   10 samples

① ② ③ ④ ⑤ . . . . . ㊿   50 samples

① ② ③ ④ ⑤ . . . . . ⑩⓪   100 samples

Mean of samples $= \bar{\bar{X}}_{samples} = \dfrac{\Sigma \bar{X}_{samples}}{n_{samples}}$

Variance of samples $= s^2_{samples} = \dfrac{\Sigma \left(\bar{X}_{sample} - \bar{\bar{X}}_{sample\ means} \right)^2}{n_{samples} - 1}$

Standard deviation of samples* $= s_{sample} = \sqrt{s^2_{sample}}$

Frequency distribution of mean plant heights from multiple samples

All mimosa plants

True mean (μ).

True variance (σ^2)

True standard deviation (σ)

True frequency distribution of plant heights in population

* The special name, standard error of the mean, is used for the standard deviation of the sample means.

Σ = sum

Practice Set 1

1. Distinguish among the following: (a) sample, sampled population, target population; (b) random sample and biased sample.

   (a)

   (b)

2. Anthony investigated the effect of herbicide runoff on the death rate of Daphnia in his grandfather's pond that was located in Grady County, Georgia. Define the sample, sampled population, and target population.

3. Pretend you are Ben and plan to conduct the experiment on the effect of a smoke-filled atmosphere on the ripening of apples. How would you define your sample, sampled population, and target population? How would you obtain your sample?

4. Both Jackie and Tony conducted an experiment on the effect of Brand X paint in preventing the rusting of iron. Jackie used a sample of 10 nails. Tony used a sample of 100 nails. Jackie reported that Brand X paint was **not** effective in preventing rusting. Tony reported that Brand X paint was effective. In whose findings would you place the greatest confidence? Why?

© 2000 by Kendall/Hunt Publishing Company, Cothron, Giese, & Rezba, *Students and Research.*

the outcome of the experiment. Frequently, research hypotheses contain an if . . . then component that predicts the effect of changing the independent variable on the dependent variable. Examples of research hypotheses include:

- Because of increased energy expenditure, forced closure mimosa plants will be shorter than nonforced closure plants.
- Wood production in trees adjacent to herbicide-treated fields will be less than wood production in trees adjacent to nonherbicide-treated fields.
- If the concentration of Chemical X is increased, then plant growth will be reduced.

After the experiment is completed, the researcher must determine whether differences between the experimental and control groups occurred by chance or reflect true differences. Scientists infer that true differences result from the experimental treatment. With controlled experiments having no underlying or concomitant variables, this inference is valid. The larger the difference between groups, the greater the probability that a true difference exists and that the inferred action of the experimental treatment is supported. To determine whether a difference is large enough to support the decision that the experimental treatment made a difference, scientists establish a statistical hypothesis and test it at a specified level of significance.

A **statistical hypothesis** refers to populations and represents the vehicle by which findings are generalized from the sample to the population. As previously mentioned, multiple samples can be drawn from a population. Because the means of these samples form a normal distribution, the means of two samples from the same population may not be identical. If samples are drawn from two different populations, one would expect the two sample means to be different. The statistician's challenge is to determine whether the means of two samples are sufficiently different to support the decision that a true difference exists and that two different populations are represented. The statistician begins by assuming that the two samples represent the same population and have identical means. Therefore, any observed difference between two sample means occurred by chance and is not significant. These assumptions are expressed through the **null hypothesis.** Several acceptable formats exist for expressing the null hypothesis in words. One option is to state that the two population means are equal; a second is to state that they are not significantly different. When the null hypothesis is expressed in mathematical symbols, the means of the two populations (1 and 2) are equated. Examples of null hypotheses are:

- The mean height of forced closure mimosa plants is not significantly different from the mean height of nonforced closure mimosa plants.

$$\mu_{Forced} = \mu_{Non-Forced}$$

- Mean wood production in trees adjacent to herbicide and nonherbicide treated fields is not significantly different.

$$\mu_{Herbicide} = \mu_{Non-Herbicide}$$

- The mean heights of plants exposed to 0 percent, 10 percent, 20 percent, and 30 percent Chemical X are equal.

$$\mu_{0\%} = \mu_{10\%} = \mu_{20\%} = \mu_{30\%}$$

LEVEL OF SIGNIFICANCE

The level of significance (α) required for statistical significance is determined by the researcher and is affected by sample size and the nature of the experiment. Common levels of significance (α) are 0.05, 0.01, and 0.001. The level of significance (for example, 0.05) communicates the probability that the researcher erred in rejecting the null hypothesis. At 0.05, the probability of error in rejecting the null hypothesis is 5/100; whereas, at 0.01 the probability is 1/100. When an error would have a profound impact, such as in drug testing, the level of significance is reduced to 0.001 or 1/1000.

Because the means of multiple samples from a population follow a normal distribution, the probability of obtaining a specific difference between the means of two samples of a given size can be determined. An excerpt from a t sampling distribution is shown in Table 11.4.

TABLE 11.4   Excerpt from a *t* Sampling Distribution

df	Probability (Level of Significance)			
	0.1	0.05	0.01	0.001
5	2.015	2.571	4.032	6.859
.				
.				
10	1.812	2.228	3.169	4.587
.				
.				
20	1.725	2.086	2.845	3.850
.				
.				
40	1.684	2.025	2.704	3.551
.				
.				
	1.645	1.960	2.576	3.291

The influence of sample size is reflected in the degrees of freedom (df) located on the left side of the excerpt from the *t* sampling distribution. Degrees of freedom represent the number of independent observations in a sample. In a sample of six plants, the mean plant height would be determined by adding the six values for plant height and dividing by the number of plants.

$$\bar{X} = \frac{50 + 60 + 30 + 50 + 50 + 60}{6} = \frac{300}{6} = 50$$

Conversely, if the mean of the six plants and the individual heights of five plants were known, the individual height of the sixth plant could be determined. The sixth value is not free to vary but is determined by the first five, therefore, only five degrees of freedom exist.

$$50 = \frac{50 + 60 + 30 + 50 + 50 + x_i}{6}$$

In a sample of n numbers with a fixed mean, the degrees of freedom are equal to n – 1. In Amy's experiment, both the experimental and control groups contained 75 mimosa plants; thus, the total degrees of freedom is 148, (75 – 1) for the experimental group plus (75 – 1) for the control group.

The influence of sample size on the expected difference between means (statistical value) can be seen from examining numerical relationships in the vertical columns of the excerpt from the *t* sampling distribution. The larger the sample (degrees of freedom), the smaller the difference between means (statistical value) required to support the decision that the means are not from the same population. Remember, scientists have more confidence in experiments with larger numbers of subjects and repeated trials. At degrees of freedom of 5, 10, 20, and 40, statistical values required for significance are 2.571, 2.228, 2.086, and 2.025, respectively, at the 0.05 level.

The influence of the level of significance (α) can be seen from examining the numerical relationships in a horizontal row of the same excerpted *t* sampling distribution. The smaller the level of significance and error rate, the larger the difference between means (statistical value) required for significance. For example, at 20 degrees of freedom, *t* values are 2.086, 2.845, and 3.850, respectively, for the 0.05, 0.01, and 0.001 levels of significance.

To make an appropriate decision regarding the null hypothesis, statisticians compare a calculated value with the table value at the appropriate degrees of freedom and level of significance (see

Table 11.5 *Sampling Distribution for* t *test* for a *t* sampling distribution). If the calculated value is smaller than the required table value, the **null hypothesis is not rejected.** When the calculated value equals or exceeds the required table value, the **null hypothesis is rejected.** For example, if a table value of 2.600 was required for significance at the 0.05 level, the following decisions would be appropriate:

Calculated value	Table value (0.5 level)	Decision about null hypothesis	Probability (p)
2.542	2.600	Not Reject	>0.05
2.600	2.600	Reject	0.05
2.785	2.600	Reject	<0.05

PROBABILITY OF ERROR

When the calculated and the table values are equal, the probability (p) of error in rejecting the null hypothesis is equal to the level of significance (α). When the calculated value is larger, the probability (p) of error is less than the level of significance (α). With detailed tables, the exact probability level can be determined by locating the probability level at which the calculated value equals the table value, for example, p = 0.001 rather than p < 0.05.

Rejection of the null hypothesis supports the alternative decision that a true difference exists between the means and that the samples represent different populations. Because a researcher generally hypothesizes a difference between means, rejection of the null hypothesis lends support to the research hypothesis. In the rare instances where a researcher hypothesizes equivalent means, nonrejection of the null hypothesis lends support to the research hypothesis.

THE *t* TEST

For quantitative data, the *t* test can be used to determine if observed differences between means of two groups are statistically significant. The *t* test is essentially a ratio that compares the difference between two means with the total standard deviation within the groups. The resulting quotient is the number of standard errors of the mean (standard deviation) between the two populations.

Several formulas exist for calculating *t*. The specific formula used depends on the number of measurements in each sample and the relationship between the measurements in the two groups being compared. When no relationship exists between measurements in the two groups, for example, experimental and control groups, an **uncorrelated *t* test** is used. Two basic formulas for calculating an uncorrelated *t* test are provided below:

Formula 1: Uncorrelated *t* Test with Equal Sample Size

$$t = \frac{\overline{X}_1 - \overline{X}_2}{\sqrt{\dfrac{s_1^2 + s_2^2}{n}}}$$

$\overline{X}_1$ = mean of Group 1
$\overline{X}_2$ = mean of Group 2
s_1^2 = variance of Group 1
s_2^2 = variance of Group 2
n = number of items or measurements in Sample 1 = Sample 2

Formula 2: Uncorrelated t Test with Unequal Sample Size

$$t = \frac{\bar{X}_1 - \bar{X}_2}{\sqrt{\dfrac{(n_1 - 1)s_1^2 + (n_2 - 1)s_2^2}{n_1 + n_2 - 2} \cdot \left(\dfrac{1}{n_1} + \dfrac{1}{n_2}\right)}}$$

$\bar{X}_1$ = mean of Group 1
$\bar{X}_1$ = mean of Group 1
$\bar{X}_2$ = mean of Group 2
s_1^2 = variance of Group 1
s_2^2 = variance of Group 2
n_1 = number of items or measurements in Sample 1
n_2 = number of items or measurements in Sample 2

When measurements in the groups are related, such as pretest and posttest data, a correlated t test is used. For use of the correlated t test and other applications, such as the significance of a difference between percents and one- and two-tailed t test, consult the statistical references listed at the end of the chapter.

TABLE 11.5 Sampling Distribution for t test

	Sampling distribution for t test			
Degrees of freedom	**Probability (Level of significance)**			
	0.1	**0.05**	**0.01**	**0.001**
1	6.314	12.706	63.657	636.619
2	2.920	4.303	9.925	31.598
3	2.353	3.182	5.841	12.924
4	2.132	2.776	4.604	8.610
5	2.015	2.571	4.032	6.864
6	1.943	2.447	3.707	5.959
7	1.895	2.365	3.499	5.408
8	1.860	2.306	3.355	5.041
9	1.833	2.262	3.250	4.781
10	1.812	2.228	3.169	4.587
11	1.796	2.201	3.106	4.437
12	1.782	2.179	3.055	4.318
13	1.771	2.160	3.012	4.221
14	1.761	2.145	2.977	4.140
15	1.753	2.131	2.947	4.073
16	1.746	2.120	2.921	4.015
17	1.740	2.110	2.898	3.965
18	1.734	2.101	2.878	3.922
19	1.729	2.093	2.861	3.883
20	1.725	2.086	2.845	3.850
21	1.721	2.080	2.831	3.819
22	1.717	2.074	2.819	3.792
23	1.714	2.069	2.807	3.767
24	1.711	2.064	2.797	3.745
25	1.708	2.060	2.787	3.725
26	1.706	2.056	2.779	3.707
27	1.703	2.052	2.771	3.690
28	1.701	2.048	2.763	3.674
29	1.699	2.045	2.756	3.659
30	1.697	2.042	2.750	3.646
40	1.684	2.021	2.704	3.551
60	1.671	2.000	2.660	3.460
120	1.658	1.980	2.617	3.373
∞	1.645	1.960	2.576	3.291

© 2000 by Kendall/Hunt Publishing Company, Cothron, Giese, & Rezba, *Students and Research*.

From STATISTICAL TABLES, 2/E by F. James Rohlf and Robert A. Sokal. Copyright © 1969, 1981 by W. H. Freeman and Company. Reprinted by permission.

Practice Set 2

1. Distinguish among the following:

    A. null hypothesis, research hypothesis
    B. degrees of freedom, sample size, level of significance
    C. calculated statistical value, sampling distribution of a statistic

2. For the following levels of significance, indicate the probability that differences occurred by chance and the probability that differences resulted from the experimental treatment: (a) 0.05, (b) 0.03, (c) 0.15, (d) 0.40, (e) 0.001.

3. Write a null hypothesis for each of the following research hypotheses:

    A. If the amount of Chemical X added to water is increased (0, 1, 2, 3 scoops), then the mean temperature of the solution will increase;
    B. Because of higher concentration of automobile pollution, loblolly pines along an interstate will exhibit a lower mean height than trees along rural roads;
    C. Lower stress levels, as exhibited by blood pressure, will occur in people who exercise;
    D. If the slope of a stream is increased (10°, 20°, 30°), then the mean sediment load will increase.

4. For each of the following degrees of freedom and levels of significance, find the statistical value required for significance (Note: Use the sampling distribution of t located in Table 11.5): (a) df = 7, α = 0.05; (b) df = 18, α = 0.01; (c) df = 30, α = 0.001; (d) df = 60, α = 0.05.

5. Several students conducted statistical tests on their experimental data. Below are the calculated statistical values and the sampling distribution values required for significance at the 0.05 level. Indicate which decision should be made by the researcher about the (a) null hypothesis—reject or not reject—and (b) research hypothesis—support or not support.

    A. Larry's calculated value = 6.954; sampling distribution value = 6.954 at 0.01 level of significance; Larry hypothesized differences.
    B. Gloria's calculated value = 8.254; at α = 0.05, df = 27, sampling distribution value = 7.954; Gloria hypothesized differences.
    C. Janet's calculated value = 10.784; sampling distribution value = 13.724 at 0.05 level of significance; Janet hypothesized no differences.

6. Using 0.05 as the maximum level required for significance, indicate what level of significance you would establish for each of the following experiments. Justify your answer.

    A. The Effect of Three Coatings on the Rusting of Iron; sample size = 100.
    B. The Effectiveness of Compound Z in Preventing Hardening of the Arteries, sample size = 500.
    C. The Rate of Molding of Bread in Plastic, Glass, Metal, and Wooden Containers; sample size = 15.

© 2000 by Kendall/Hunt Publishing Company, Cothron, Giese, & Rezba, *Students and Research*.

TWO TREATMENT GROUPS

Scenario: John read that farmers in Japan routinely subject plants to stress before transplanting from the greenhouse to the field. Methods of stress induction included pulling on the plants and hitting them with straw rakes. John decided to investigate this phenomenon by growing two groups of bean plants (10/group) in a greenhouse for 15 days during which time the plants in one group were pulled on three times daily at 8:00 in the morning and at 4:00 in the afternoon. The plants were then transplanted to a field. John hypothesized that stressed plants would exhibit greater mean height after transplanting than the nonstressed plants (control). Plant heights after 30 days follow.

Stressed Plants	Nonstressed Plants
55.0	48.0
65.0	65.0
50.0	59.0
57.0	57.0
59.0	51.0
73.0	63.0
57.0	65.0
54.0	58.0
62.0	44.0
68.0	50.0

Step 1: State the null hypothesis.

Null Hypothesis: The mean height of stressed plants is **not significantly different** from the mean height of nonstressed plants.

H_0: $\mu_{stressed} = \mu_{nonstressed}$

Step 2: Establish the level of significance.

Alpha (α) = 0.05

Step 3: Calculate the means.

$$\overline{X}_{stressed} = \frac{\Sigma X_i}{n}$$

$$= \frac{55 + 65 + 50 + 57 + 59 + 73 + 57 + 54 + 62 + 68}{10} = \frac{600}{10}$$

$$= 60.0$$

$$\overline{X}_{nonstressed} = \frac{\Sigma X_i}{n}$$

$$= \frac{48 + 65 + 59 + 57 + 51 + 63 + 65 + 58 + 44 + 50}{10} = \frac{560}{10}$$

$$= 56.0$$

Step 4:   Calculate the variance.

$$s^2_{stressed} = \frac{\Sigma(X_i - \overline{X})^2}{n - 1}$$

$$= \frac{(55 - 60)^2 + (65 - 60)^2 + (50 - 60)^2 \ldots + (68 - 60)^2}{10 - 1}$$

$$= 49.1$$

$$s^2_{nonstressed} = \frac{\Sigma(X_i - \overline{X})^2}{n - 1}$$

$$= \frac{(48 - 56)^2 + (65 - 56)^2 + (59 - 56)^2 \ldots + (50 - 56)^2}{10 - 1}$$

$$= 54.9$$

Step 5:   Calculate t.

$$t = \frac{\overline{X}_1 - \overline{X}_2}{\sqrt{\dfrac{s_1^2 + s_2^2}{n}}}$$

$$t = \frac{60 - 56}{\sqrt{\dfrac{49.1 + 54.9}{10}}} = \frac{4}{\sqrt{\dfrac{104.0}{10}}} = \frac{4}{\sqrt{10.4}} = \frac{4}{3.2}$$

$$t = 1.25 \sim 1.3$$

Step 6:   Determine the degrees of freedom.

df   = (Number of stressed plants − 1)
        + (Number of nonstressed plants − 1)
df   = (10 − 1) + (10 − 1)
df   = 9 + 9 = 18

Step 7:   Determine the significance of the calculated t.

At df = 18, α at 0.05, t = 2.101; the calculated t of 1.3 < 2.101 and is **not significant** at the 0.05 level (p > 0.10).

Step 8:   Decide to reject or not reject the null hypothesis.
Because the calculated value of t is **not significant,** the null hypothesis is **not rejected.**

$$H_0: \mu_{stressed} = \mu_{nonstressed}$$

Step 9:   Determine whether the statistical findings support the research hypothesis.
Because the null hypothesis was **not rejected** at the 0.05 level of significance, the research hypothesis that stressed plants would have a greater mean height than nonstressed plants was **not supported.**

Step 10: Construct a data table that communicates both descriptive and inferential statistics.

In Chapter 8, a model for constructing a quantitative data table that included descriptive information on the sample (mean, variance, and standard deviation) is described. Simply expand the table to include the results of the test as depicted in Table 11.6 *Effect of Stress on the Mean Height of Bean Plants.*

TABLE 11.6   Effect of Stress on the Mean Height of Bean Plants

Descriptive information	Stressed group	Nonstressed group
Mean	60.0	56.0
Variance	49.1	54.9
Standard deviation	7.0	7.4
1 SD (68% Band)	53.0 – 67.0	48.6 – 63.4
2 SD (95% Band)	46.0 – 74.0	41.2 – 70.8
3 SD (99% Band)	39.0 – 81.0	33.8 – 78.2
Number	10	10
Results of *t* test	$t = 1.3$ df $= 18$ *t* of $1.3 < 2.101$	$\alpha = 0.05$ $p > 0.10$

Step 11: Write a paragraph describing results.

Chapter 9 provides a structured process for writing about data. Using a similar technique, show students how to write sentences that communicate critical information.

Steps	*Examples*
1. Write a topic sentence stating the independent and dependent variables, and a reference to tables or graphs.	Effects of stress on the height of bean plants are summarized in Table 11.6.
2. Write sentences comparing the measures of central tendency (means) and variation (standard deviation) of the groups.	Stressed plants exhibited a greater mean height (60 cm) than nonstressed plants (56.0 cm). Variations within the groups were similar, with stressed plants having a standard deviation of 7.0 and nonstressed plants a standard deviation of 7.4. Ninety-five percent of the stressed plants fell within the range of 46.0 to 74.0 cm, as opposed to nonstressed plants, which ranged from 41.2 to 70.8 cm.
3. Write sentences describing the statistical test, level of significance, and null hypothesis.	The uncorrelated *t* test was used to test the following null hypothesis at the 0.05 level of significance: The mean height of stressed plants is not significantly different from the mean height of non-stressed plants.

Steps	*Examples*
4. Write sentences comparing the calculated value with the required statistical values and make a statement about rejection of the null hypothesis.	The null hypothesis was not rejected ($t = 1.3 < 2.101$ at df $= 18$; p. > 0.10)
5. Write sentences stating support of the research hypothesis by the data.	The data did not support the research hypothesis that stressed plants would have a different mean height after planting than nonstressed plants.

Step 12: Write an appropriate conclusion.

A structured process for writing conclusions is also provided in Chapter 9. Using the following questions as a guideline, show students how to write a concluding paragraph that includes the results of the statistical test.

Questions	*Examples*
1. What was the purpose of the experiment?	The effect of stress on the growth of bean plants was investigated by comparing the growth of bean plants subjected to stress for 15 days with a control (nonstressed plants).
2. What were the major findings? (Focus on results of the statistical test).	No significant difference existed between the mean height of stressed plants and nonstressed plants 30 days after transplanting.
3. Was the research hypothesis supported by the data?	The research hypothesis that stressed plants would have a different mean height was not supported.
4. How did your findings compare with those of other researchers?	In contrast, Japanese farmers found that hitting and pulling rice plants were beneficial.
5. What possible explanations can you offer for your findings?	Possible explanations include differences in the methods of administering stress or the type of plant, for example, monocots (rice) versus dicots (beans).
6. What recommendations do you have for further study and for improving the experiment?	Additional investigations using various sources of stress at more frequent intervals with both monocots and dicots should be conducted. Improved experimental design techniques should be implemented, including a larger sample size and a longer growing period.

THREE TREATMENT GROUPS

Because the *t* test allows only for the comparison of two means, multiple *t* tests would be necessary if more than two experimental groups were involved. For example, if an experiment involved three treatment groups (A, B, C), the number of possible *t* tests would be three (A = B; A = C; B = C). With four and five treatment groups, the number of *t* tests would increase to 6 and 10, respectively. The probability of obtaining a significant *t* value by chance increases with the number of *t* tests conducted. Additionally, interactions exist among groups. For these reasons, multiple *t* tests are not recommended if options exist for using more sophisticated statistical tests such as analysis of variance. With secondary students, such options do not always exist. Many judges of competitions are lenient because they realize that the concept of a statistical test is of primary importance, rather than the sophistication of the test. However, problems with multiple *t* tests should be recognized and appropriate actions taken to minimize obtaining significant differences by chance. These include changing the level of significance (from 0.05 to 0.01) and conducting the minimum number of *t* tests necessary to ascertain support for the research hypotheses. For example, if a student hypothesized that 10 percent and 20 percent Chemical X would promote plant growth while 30 percent X would retard growth, as compared with a control, only 3 *t* tests, not 6, would be required.

Required for Testing Hypotheses	Maximum That Could Be Conducted
10% = 0%	10% = 0%
20% = 0%	10% = 20%
30% = 0%	10% = 30%
	20% = 0%
	20% = 30%
	30% = 0%

The following experiment illustrates a typical use of multiple *t* tests by high school students.

Scenario: Laura read about recycling of plant materials (grass and leaves) and their use as fertilizer by making a compost pile. Laura decided to investigate the relative effectiveness of commercial fertilizer and compost in promoting the growth of radish plants. Laura hypothesized that plants grown with compost and plants grown with commercial fertilizer would exhibit similar heights and that both fertilized groups would exhibit greater heights than the control. Plant heights (cm) after three weeks of growth were as follows:

Commercial Fertilizer	Compost Fertilizer	No Fertilizer (Control)
5.0	5.0	3.0
6.0	6.0	5.0
10.0	4.0	2.0
6.0	7.0	3.0
8.0	3.0	4.0
6.0	3.0	6.0
5.0	4.0	6.0
8.0	7.0	3.0
6.0	6.0	5.0
10.0	5.0	3.0

Step 1: State the null hypotheses.

$1H_0$: $\mu_{compost}$ = $\mu_{commercial}$
$2H_0$: $\mu_{compost}$ = $\mu_{control}$
$3H_0$: $\mu_{commercial}$ = $\mu_{control}$

Step 2: Establish the level of significance.

Alpha $(\alpha) = 0.01$

Step 3: Calculate the means.

$$\overline{X}_{commercial} = \frac{5 + 6 + 10 \ldots + 10}{10} = 7$$

$$\overline{X}_{compost} = \frac{5 + 6 + 4 \ldots + 5}{10} = 5$$

$$\overline{X}_{control} = \frac{3 + 5 + 2 \ldots + 3}{10} = 4$$

Step 4: Calculate the variance.

$$s^2_{commercial} = \frac{\Sigma(X_i - \overline{X})^2}{n - 1} = 3.6$$

$$s^2_{compost} = \frac{(5 - 5)^2 + (6 - 5)^2 + (4 - 6)^2 + \ldots (5 - 6)^2}{10 - 1}$$
$$= 2.2$$

$$s^2_{control} = \frac{(3 - 4)^2 + (5 - 4)^2 + \ldots (5 - 4)^2 + (3 - 4)^2}{10 - 1}$$
$$= 2.0$$

Step 5: Calculate t.

$$t = \frac{\overline{X}_1 - \overline{X}_2}{\sqrt{\dfrac{s^2_1 + s^2_2}{n}}}$$

$$t_{commercial\ vs\ compost} = \frac{7.0 - 5.0}{\sqrt{\dfrac{3.6 + 2.2}{10}}} = 2.62 \sim 2.6$$

$$t_{compost\ vs\ control} = \frac{5.0 - 4.0}{\sqrt{\dfrac{2.2 + 2.0}{10}}} = 1.51 \sim 1.5$$

$$t_{commercial\ vs\ control} = \frac{7.0 - 4.0}{\sqrt{\dfrac{3.6 + 2.0}{10}}} = 4.01 \sim 4.0$$

Step 6:   Determine the degrees of freedom.

df = (Number of plants in commercial − 1) + (Number of plants in compost − 1);
df = 9 + 9 = 18.

Similarly, the degrees of freedom are 18 for the other two t tests.

Step 7:   Determine the significance of the calculated t.

At df = 18, α of 0.01, t = 2.878; as outlined below, only the calculated t for the commercial-control comparison is significant at the 0.01 level.

$t_{commercial\ vs\ compost}$ = 2.6 < 2.878; 0.01 < p < 0.05 not significant
$t_{compost\ vs\ control}$ = 1.5 < 2.878; p > 0.01   not significant
$t_{commercial\ vs\ control}$ = 4.0 > 2.878; p < 0.001   significant

Step 8:   **Decide whether to reject the null hypotheses.**
Because the calculated values of t are not significant for the commercial-compost and compost-control comparisons, the null hypotheses are **not rejected.** For the commercial-control comparison, the calculated t is significant; thus, the null hypothesis is **rejected.**

Step 9:   **Determine if the statistical findings support/do not support the research hypothesis.**
Results of the uncorrelated t test support the first research hypothesis that plants grown with commercial and compost fertilizer would exhibit similar heights. The hypothesized greater growth of plants fertilized with commercial fertilizer, as compared with a control, was also supported by the data. Because no significant differences were found between the height of plants fertilized with compost and the control, the research hypothesis of increased growth with compost was not supported.

Step 10: **Construct a data table** (see Table 11.7 *Effect of Fertilizer on the Mean Height (cm) of Bean Plants*).

TABLE 11.7   Effect of Fertilizer on the Mean Height (cm) of Bean Plants

Descriptive information	Commercial	Compost	Control
Mean	7.0	5.0	4.0
Variance	3.6	2.2	2.0
Standard deviation	1.9	1.5	1.4
1 SD (68% Band)	5.1 – 8.9	3.5 – 6.5	2.6 – 5.4
2 SD (95% Band)	3.2 – 10.8	2.0 – 8.0	1.2 – 6.8
3 SD (99% Band)	1.3 – 12.7	0.5 – 9.5	0 – 8.2
Number	10	10	10
Results of t test			
Commercial vs Compost t = 2.6; .01 < p < 0.05			
Compost vs Control t = 1.5; p > 0.01			
Commercial vs Control t = 4.0; p < 0.001			
At df of 18; α of 0.01; t = 2.878 for significance			

Step 11: Write a paragraph describing results.

The relative effects of commercial fertilizer and compost on the growth of bean plants, as compared with a control group, are summarized in Table 11.7. The mean height (7.0 cm) of plants grown with commercial fertilizer was greater than the mean height of plants grown with compost (5.0 cm) or the control (4.0 cm). Greater variation in plant height was found in the commercial fertilizer group, standard deviation of 1.9, than in the compost group (SD = 1.5) or the control (SD = 1.4). The uncorrelated t test was used to test three null hypotheses at the 0.01 level of significance.

Null Hypothesis 1: The mean height of plants fertilized with compost is not significantly different from the mean height of plants fertilized with commercial fertilizer.

Null Hypothesis 2: The mean height of plants fertilized with compost is not significantly different from the mean height of plants receiving no fertilizer (control).

Null Hypothesis 3: The mean height of plants fertilized with commercial fertilizer is not significantly different from the mean height of plants receiving no fertilizer (control).

Null Hypothesis 1 was not rejected; the mean height of plants fertilized with commercial fertilizer was not significantly greater than the mean height of plants fertilized with compost ($t = 2.62 < 2.878$ at df = 18; $0.01 < p < 0.05$). Null Hypothesis 2 was not rejected, because the mean heights of plants receiving compost and the control were equivalent ($t = 1.51 < 2.878$ at df = 18; $p > 0.01$). Null Hypothesis 3 was rejected, with plants receiving commercial fertilizer exhibiting greater mean height than the control group ($t = 4.01 > 2.878$ at df = 18; $p < 0.01$). These findings support the research hypotheses that commercial fertilizer and compost would have similar effects on plant growth and that commercial fertilizer would be superior to the control. The data do not support the hypothesis that compost would promote plant growth compared with a control.

Step 12: Write an appropriate conclusion.

The relative effectiveness of commercial fertilizer and compost in promoting radish growth was investigated. Mean height of plants receiving commercial fertilizer was significantly greater than the control. No significant differences in heights were found

© 1998 PhotoDisc, Inc.

between plants receiving compost and commercial fertilizer. The findings support the hypothesized equivalency of commercial and compost fertilizer and the superiority of commercial fertilizer, as compared with a control. The hypothesized superiority of compost as compared with a control was not supported by the data. Black (1999), Stone (1997), and Jones (1998) reported superior results with compost. Age of compost in these experiments differed and a longer growing period (45–60 days) was used. Further experimentation is necessary to determine if other types of compost of different ages are beneficial to plant growth over an extended period.

Practice Set 3

1. Gail Adams investigated the effect of time of application of an herbicide on the yield of soybeans. Gail hypothesized that herbicides applied early in the morning would be more effective than at midday. Determine if Gail's hypothesis was supported.

	8:00 A.M.	12:00 P.M.
Mean Yield *(g)*	70.8	77.4
Standard deviation	30.8	31.2
Number *(n)*	30	30

2. Using Amy's data from the introduction to this chapter, determine whether the mean height of the forced-closure mimosa seedlings differed significantly from the control group.

3. Under the supervision of a mentor, David Setchel investigated the effect of testosterone and estrogen on the prostate weight (mg/100 g body mass) of rats. Descriptive information on the samples was provided in his paper:

	Testosterone	Estrogen	Control
Mean	70.34	53.55	53.16
Variance	199.34	190.35	57.27
Standard deviation	14.14	13.80	7.57
Number *(n)*	20	20	20

    Determine if David's research hypothesis that prostate mass would be positively affected by testosterone and negatively affected by estrogen was supported by the data.

(continued on the following page)

Practice Set 3 (continued)

4. In the practice problems in Chapter 10, Ms. Goldfarb's students thought that sometimes they didn't get their money's worth at a local fast food restaurant when they bought a jumbo order of french fries. Different servers seemed to vary a lot in the number of french fries they put in each order. The students designed a study to see if this was true, and on a particular day, they asked people to count the number of french fries they were served.

    A. Use a *t* test to determine if there is a statistically significant difference in the numbers of french fries each server served.
    B. Between which servers do you think there is a statistically significant difference?

Servers / Test	*t* value	p value	Significant Y/N?
1 & 2			
1 & 3			
1 & 4			
2 & 3			
2 & 4			
3 & 4			

5. Conduct Investigation 11.1 *Magnetic Time*. Use a *t* test to analyze your data. Communicate your findings through an appropriate data table, graph, and paragraph.

INVESTIGATION 11.1 • Magnetic Time

Question

How does separation method affect time to separate a mixture?

Mixture	Magnetic

Materials for Each Group of Two Students

● Magnet

● White paper, letter size (8.5 in. x 11.0 in.)

● Plastic zip bag (pint size) with mixture of magnetic and non-magnetic objects. Examples include paper clips and pinto beans or metal paper clips and non-magnetic (brass) paper fasteners or plastic and metal paper clips.

● Watch or clock with second hand.

Procedure

1. Draw a line to divide a sheet of notebook sized paper in half. Label one side "mixture" and the other side "magnetic."

2. Pour the contents of the zip bag on the side labeled "mixture." Be sure all of the items are on that side.

3. By hand, sort the magnetic objects into the other side labeled "magnetic." Record the time (sec).

4. Return the objects to the zip bag, close, and shake to evenly distribute items.

5. Repeat Steps 2–4 using a magnet to separate the items. Record the time (sec).

6. Have your partner do Steps 1–5.

7. Record your data and your partner's in a class data table. Be sure that at least 30 sets of data are entered. If there are fewer than 30 trials, then some students should repeat Steps 1–6 to obtain sufficient trials.

Analyzing Your Data

1. Make a data table to display the class data. Use Table 11.4 and/or 11.5 as a model.

2. Conduct a *t* test to determine if there is a significant difference in the time required for sorting the objects by hand and with a magnet. Enter the appropriate information into the data table.

3. Write a paragraph of results. (Hint: See paragraph accompanying Table 11.4 and/or 11.5.)

(continued on the following page)

INVESTIGATION 11.1 • Magnetic Time (continued)

USING TECHNOLOGY

1. In the **STAT** mode of your calculator, enter the time to separate by hand in List 1 and the time to separate with a magnet in List 2. (See Appendix A, *Using Technology*, for additional help in using the graphing calculator).

2. Obtain 1-variable statistics for each set of data, such as the mean, standard deviation, minimum, maximum, and number. In the **STAT** mode select CALC (for calculate) and then 1 VAR (for 1-variable statistics). Depending on the brand of your calculator, you will need to enter the desired list number (e.g., L1), or SET the 1-variable x-list to the desired list number before selecting 1 VAR. *Repeat* the selection process for the second set of data by changing the list number.

3. Then select **TESTS** and choose the 2-sample *t* test.

4. Enter List 1 and List 2 as the locations of the data, select *Yes* for pooled data, and then calculate.

5. Among the values provided are the *t* value, the probability the differences are due to chance, the degrees of freedom, and the means for both data sets.

Extending Your Learning

1. How are magnets used in industry, manufacturing, and mining to separate objects?

2. What affects the strength of a magnet? Do you think you would get the same results with magnets of different sizes, ages, compositions? How could you design experiments to test your hypothesis? What other ways could you measure the dependent variable?

3. Do you think that the type of mixture affects the task? What other mixtures could you use? How could you design experiments to test your hypothesis?

4. What other factors might affect the outcome of this task?

CHI-SQUARE

For qualitative data, chi-square (χ^2) can be used to determine if differences between frequency distributions are statistically significant. An observed frequency distribution may be compared with an expected or theoretical frequency distribution for goodness of fit using the following formula.

$$\chi^2 = \sum \frac{(O - E)^2}{E}$$

χ^2 = Chi-square
Σ = Sum of the Values
O = Observed Frequency Distribution
E = Expected Frequency Distribution

The calculated χ^2 value is compared with a sampling distribution of χ^2 to determine significance (see Table 11.8 *Chi-Square Sampling Distribution*). Chi-square may also be used to evaluate whether two variables are associated or related; for use of such contingency tables, consult the statistical references listed at the end of the chapter.

GOODNESS OF FIT

Scenario: Mary read that bees were attracted to the color yellow as opposed to red, blue, or white. She wondered if crickets would show a color preference. To test her hypothesis that crickets would be differentially attracted to colors, she placed 100 crickets in a container. The bottom of the container was divided into four equal sections covered by red, blue, yellow, or white paper. She observed the number of crickets on each color paper one hour after placing them in the container. The distribution of crickets was: 30 red, 40 blue, 12 yellow, and 18 white. By chance alone, an equal number of crickets on each color of paper would be expected.

Step 1: **State the null hypothesis.**

Null Hypothesis: The frequency distribution of crickets on various colors is **not significantly different** from the frequency distribution predicted by chance.

Observed Frequency Distribution on Various Colors = Expected (Chance) Frequency Distribution on Various Colors

Step 2: **Establish the level of significance.**

Alpha (α) = 0.05

Step 3: **Determine the observed frequency distribution.**

		Red	Blue	Yellow	White
Overall Distribution	=	30	40	12	18

Step 4: **Determine the expected theoretical frequency distribution.**
By chance, one would expect an equal distribution of crickets across the four colors. Because a total of 100 crickets was used, 1/4 of 100 or 25 crickets are expected per color.

		Red	Blue	Yellow	White
Expected Distribution	=	25	25	25	25

TABLE 11.8   Chi-Square Sampling Distribution

Degrees of Freedom	Probability (Level of significance)			
	0.1	0.05	0.01	0.001
1	2.706	3.841	6.635	7.879
2	4.605	5.991	9.210	10.597
3	6.251	7.815	11.345	12.830
4	7.779	9.488	13.277	14.860
5	9.236	11.070	15.086	16.750
6	10.645	12.592	16.812	18.548
7	12.017	14.067	18.475	20.278
8	13.362	15.507	20.090	21.955
9	14.684	16.919	21.666	23.589
10	15.987	18.307	23.209	25.188
11	17.275	19.675	24.725	26.757
12	18.549	21.026	26.217	28.300
13	19.812	22.362	27.688	29.819
14	21.064	23.685	29.141	31.319
15	22.309	24.996	30.578	32.801
16	23.542	26.296	32.000	34.267
17	24.769	27.587	33.409	35.718
18	25.989	28.869	34.805	37.156
19	27.204	30.144	36.191	38.582
20	28.412	31.410	37.566	39.997
21	29.615	32.670	38.932	41.401
22	30.813	33.924	40.289	42.796
23	32.007	35.172	41.638	44.181
24	33.196	36.415	42.980	45.558
25	34.382	37.652	44.314	46.928
26	35.563	38.885	45.642	48.290
27	36.741	40.113	46.963	49.645
28	37.916	41.337	48.278	50.993
29	39.088	42.559	49.588	52.336
30	40.256	43.773	50.892	53.672

© 2000 by Kendall/Hunt Publishing Company, Cothron, Giese, & Rezba, *Students and Research*.

From STATISTICAL TABLES, 2/E by F. James Rohlf and Robert A. Sokal. Copyright © 1969, 1981 by W. H. Freeman and Company. Reprinted by permission.

Step 5:   Calculate χ^2.

$$\chi^2 = \frac{(30-25)^2}{25} + \frac{(40-25)^2}{25} + \frac{(12-25)^2}{25} + \frac{(18-25)^2}{25}$$

$$= \frac{(5)^2}{25} + \frac{(15)^2}{25} + \frac{(13)^2}{25} + \frac{(7)^2}{25}$$

$$= \frac{25}{25} + \frac{225}{25} + \frac{169}{25} + \frac{49}{25}$$

$$= 1.0 + 9.0 + 6.7 + 1.9$$

$$= 18.6$$

Step 6:   Determine the degrees of freedom.

df = (Number of Categories – 1)
df = (4 – 1)
df = 3

Step 7:   Determine the significance of the calculated χ^2.

At df = 3, α of 0.05, χ^2 = 7.815 for significance; the calculated χ^2 of 18.6 > 7.815 and is significant at the 0.05 level; $p < 0.001$.

Step 8:   **Decide to reject or not reject the null hypothesis.**
Because the calculated χ^2 is significant, the null hypothesis is rejected.

Step 9:   **Determine whether the statistical findings support the research hypothesis:**
Because the null hypothesis was rejected, the research hypothesis that crickets would be differentially attracted to colors was supported.

Step 10: **Construct a data table** (see Table 11.9 *Attraction of Crickets to Various Colors*).

TABLE 11.9   Attraction of Crickets to Various Colors

Information	Observed distribution	Expected distribution (Chance)	Calculated χ^2
Mode	Blue	Red-Blue-Yellow-White	
Frequency distribution			
Red	30	25	1.0
Blue	40	25	9.0
Yellow	12	25	6.7
White	18	25	1.9
Number	100	100	
Results of the chi-square test	χ^2 = 18.16 at df = 3 χ^2 of 18.6 > 7.815 $p < 0.001$		

Step 11: Write a paragraph describing the results.

The distribution of crickets on various colors, as compared with a chance distribution is summarized in Table 11.9. More crickets appeared on red and blue and fewer on yellow and white than was predicted by chance. Chi-Square was used to test the following null hypothesis at the 0.05 level of significance:

The frequency distribution of crickets on various colors is not significantly different from the frequency distribution predicted by chance.

The null hypothesis was rejected ($\chi^2 = 18.6 > 7.815$ at df = 3; $p < 0.001$). The data supported the research hypothesis that crickets were differentially attracted to the color blue and repelled by the color yellow.

Step 12: Write an appropriate conclusion.

The relative attraction of crickets to the colors red, yellow, blue, and white was investigated by comparing the distribution of 100 crickets on the colors with a chance distribution. Significant differences existed, with crickets being repelled by the color yellow and attracted to the color blue. Although the research hypothesis was supported, the findings conflicted with the documented attraction of bees to yellow. Additional investigations should be conducted to determine if the frequency of the color used was a factor or if true interspecies differences occur. Adaptive benefits of the color preferences could also be determined including food procurement or prey avoidance.

Multiple chi-square comparisons must be employed when more than one experimental group is involved in an experiment. As previously described for the *t* test, the probability of obtaining a significant value by chance increases with the number of comparisons. Therefore, chi-square tests should be reduced to the minimum needed to answer the research question, and the level of significance should be increased to 0.01 or above. Chi-Square tests should not be conducted when the observed frequency count is less than 5; categories may be combined to avoid these problems.

MULTIPLE COMPARISONS

Scenario: Steven read that the ratio of black to white moths in England increased over time as the environment became covered with soot. He derived a simulation for testing the effect of environmental changes on the type of offspring surviving. Newspaper was used as the background (environment) from which three types of paper prey (red, white, and newsprint) were randomly selected. Initially, 99 prey, 33 of each color, were placed upon the newsprint. Simulating a predator, Steven randomly selected 9 prey from the environment; remaining organisms were doubled to simulate reproduction and the selection process repeated. Prey of each color were counted and the relative number of each based upon 100 were expressed. The activity was repeated for the equivalent of four generations. Steven hypothesized that differential prey distribution would occur by the third generation. He used chi-square to determine if a significant difference in the distribution of prey occurred between Generation 0 and 1, Generation 0 and 2, Generation 0 and 3.

	Red prey	White prey	Newsprint prey
Generation 0	33.0	33.0	33.0
Generation 1	32.0	30.0	38.0
Generation 2	27.0	25.0	48.0
Generation 3	15.0	20.0	65.0

Step 1: **Null hypotheses:**

$1H_0$: Frequency Distribution = Frequency Distribution
Moths Generation 0   Moths Generation 1

$2H_{20}$: Frequency Distribution = Frequency Distribution
Moths Generation 0   Moths Generation 2

$3H_0$: Frequency Distribution = Frequency Distribution
Moths Generation 0   Moths Generation 3

Step 2: Alpha $(\alpha) = 0.01$.

Step 3: The observed frequency distributions for Generations 1, 2, and 3 are stated in the scenario.

Step 4: The expected (theoretical) frequency distribution is equivalent to Generation 0, as stated in the scenario.

Step 5: Calculate χ^2.

$$\chi^2_{0-1} = \frac{(32-33)^2}{33} + \frac{(30-33)^2}{33} + \frac{(38-33)^2}{33}$$

$$= .03 + .27 + .76$$
$$= 1.05$$

$$\chi^2_{0-2} = \frac{(27-33)^2}{33} + \frac{(25-33)^2}{33} + \frac{(48-33)^2}{33}$$

$$= 1.09 + 1.94 + 6.82$$
$$= 9.85$$

$$\chi^2_{0-3} = \frac{(15-33)^2}{33} + \frac{(20-33)^2}{33} + \frac{(65-33)^2}{33}$$

$$= 9.82 + 5.12 + 31.0$$
$$= 45.9$$

Step 6: Degrees of Freedom $= (3 - 1) = 2$.

Step 7: Because at df = 2; α of 0.01, $\chi^2 = 9.210$ for significance, the chi-square for the Generation 0–1 comparison is not significant. The chi-square for the Generation 0–2 and Generation 0–3 comparisons are significant.

Step 8: Do not reject
$1H_0$: Frequency Distribution = Frequency Distribution
Moths Generation 0   Moths in Generation 1

Reject
$2H_0$: Frequency Distribution = Frequency Distribution
Moths Generation 0   Moths in Generation 2

Reject
$3H_0$: Frequency Distribution = Frequency Distribution
Moths Generation 0   Moths in Generation 3

Step 9: The research hypothesis that differential prey distribution would occur by the third generation was supported. Significant differences occurred by the second generation.

Steps 10–12: Using the strategies previously described, design an appropriate data table and write appropriate results and a conclusion for Steven's study.

EVALUATING AND REINFORCING SKILLS

Initially, students will need practice with pre-collected sets of data and data from laboratory experiments. Several practice exercises and paper-pencil test items are provided in Chapters 11 and 15. Students experience the greatest difficulty in identifying the appropriate statistical test for a given set of data. These skills are enhanced by requiring students to read written reports or papers from scientific research competitions and to evaluate their statistical analysis. In Chapter 16, the criteria in *Evaluating for Success 8: Analyzing and Communicating Data: Inferential Statistics* can be used to analyze these written reports as well as students' original products.

REFERENCES

Adams, G.L. *The effect of Blazer 2L herbicide on soybean yield and the amount of broadleaf weeds when applied at different times.* Paper presented at the meeting of the Virginia Junior Academy of Science, First Place, Agriculture and Animal Science, Norfolk, VA: Old Dominion University.

Glenberg, A. (1996). *Learning from data: An introduction to statistical reasoning* (2nd ed.). Mahwah, NJ: Lawrence Erlbaum Associates.

Hubbard, Brian K. *The effect of different compositions of superconductors on levitation height and resistance.* Paper presented at the meeting of the Virginia Junior Academy of Science, First Place, Physics, Richmond, VA: Virginia Commonwealth University.

Landwehr, J.M. & Watkins, A.E. (1994). *Exploring data.* A component of the Quantitative Literacy Series. Palo Alto, CA: Dale Seymour Publications.

Landwehr, J.M., Swift, J., & Watkins, A.E. (1998). *Exploring surveys and information from samples.* A component of the Data-Driven Mathematics Series. Palo Alto, CA: Dale Seymour Publications.

McClave, J.T., Dietrich, Frank H. II, & Sincich, T. (1997). *Statistics* (7th ed.). Upper Saddle River, NJ: Prentice-Hall, Inc.

Setchel, D. P. *The effects of steroid hormones and castration on the prostrate weight in male rats.* Paper presented at the meeting of the Virginia Junior Academy of Science, First Place, Zoology, Williamsburg, VA: The College of William and Mary.

Shavelson, R.J. (1996). *Statistical reasoning for the behavioral sciences* (3rd ed.). Boston: Allyn & Bacon.

Statistics Canada (www.statcan.ca)

Statistical Resources on the Web (www.lib.umich.edu/libhome/Documents.center/stats.html)

U.S. and International government statistics (www.bls.gov/oreother.htm)

Worsham, K.A. *Analysis of the external and internal sediments of the articulated fossil. Dosinia acetabulum, found in Hampton, Virginia.* Paper presented at the meeting of the Virginia Junior Academy of Science, Second Place, Geology and Earth Science, Harrisonburg, VA: James Madison University.

Wright, A. L. *The effect of sensitivity on the growth of Minosa pudica.* Paper presented at the meeting of the Virginia Junior Academy of Science, First Place, Botany, Charlottesville, VA: The University of Virginia.

Yates, D.S., Moore, O.S., McCabe, G.P. (1999). *The Practice of Statistics: TI-83 Graphing Calculator Enhanced.* New York; W.H. Freeman and Company.

Related Web Site

http://www.mste.uiuc.edu/stat/stat.html

Practice Set 4

1. Kyle Worsham investigated the effect of how open the shells of fossil clams are on the distribution of microfossils within the clam as compared with the external environment. He determined the mass (mg) of sediments inside and outside a group of 10 fossils having a valve separation of 0–10 mm. The process was repeated for fossils having a valve separation of 11–20 mm and for a group having a valve separation greater than 20 mm. Kyle hypothesized that significant differences in sediment distribution would occur when the valve separation was less than 20 mm. Analyze the data collected by Kyle to determine whether his research hypotheses are supported.

Hint: Use the internal distribution as the observed distribution and the external distribution as the expected.

Size of Shell Opening	Fraction Size						
	2 mm	*1 mm*	*.5 mm*	*.25 mm*	*.125 mm*	*.075 mm*	*<.075 mm*
0–10 mm							
Int.	2	8	14	69	560	290	54
Ext.	9	15	29	95	540	260	56
11–20 mm							
Int.	8	10	25	91	586	216	55
Ext.	14	17	35	118	565	198	52
20 mm							
Int.	20	22	37	102	557	212	51
Ext.	20	22	49	136	553	180	48

2. Conduct Investigation 11.2 *Containing the Curdles.* Use a chi-square test to analyze your data. Communicate your findings with an appropriate data table, graph, and paragraph.

INVESTIGATION 11.2 • Containing the Curdles

Question

How does temperature affect the curdling of milk?

Materials for Each Group of Two Students

150 ml of milk at room temperature
3 small clear zip bags—snack size
Ice—crushed
Beaker of hot water from faucet
3 bowls or other containers
1 graduated cylinder (100 ml)
1 graduated cylinder (10 ml) or calibrated dropper
18 ml of white vinegar
Tape
Pen or pencil

Cold (ice)

Hot (faucet)

Room temperature (empty)

Safety

● Be careful handling the hot water. The temperature of water from faucets is very different.
● Wear goggles.
● Dispose of zip bags as directed by your teacher.
● Do NOT eat/drink the foods used in this experiment.
● Wash your hands after handling food/chemicals.

Procedure

1. Using tape, label each zip bag with one of the following labels: cold, room temperature, hot.
2. Add 50 ml of milk to each zip bag. Close it tightly.
3. Place the first bag of milk in a bowl and surround the bag with crushed ice. Place the second bag of milk in an empty bowl. Place the third bag of milk in a bowl and fill the bowl with hot water from the faucet. Leave the bags of milk in the bowl for 10 min.
4. Remove the bags from the bowls. Quickly unzip each bag and add 6 ml of vinegar to each bag. Reseal and return to the appropriate bowls. Describe the appearance of the curdles using the following symbols:

   SC—small curdles with milk generally not separated and still white in color
   MC—moderate curdles with milk beginning to separate and show yellow color
   LC—large curdles with milk separated and showing yellow color

   Note: To reduce odor, do not unzip the bags after you have added vinegar. Dispose of the bags as directed by your teacher.

(continued on the following page)

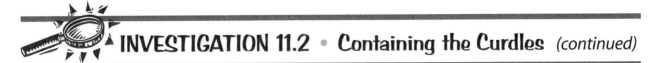

INVESTIGATION 11.2 • Containing the Curdles *(continued)*

5. Enter the data in a class data table. Compile data from several classes, so that you have 50 or more trials.

Analyzing Your Data

1. Conduct a chi-square test to determine if significant differences exist among the appearance of the milk curdles at various temperatures, e.g. room vs. cold, room vs. hot, cold vs. hot.
2. Summarize your results with an appropriate data table, graph, and paragraph. (Hint: See Table 11.9 and accompanying paragraph.)
3. What is the chemical composition of milk? What causes the souring of milk? What factors affect the speed at which this chemical reaction occurs?
4. Do you think that you would get the same results with all types of milk? With various types of acids? How could you design experiments to test your hypotheses?

USING TECHNOLOGY ·

*When you have **one** independent variable, as in this investigation:*

Because graphing calculators only include a built-in program for a more complex chi-square test for two independent variables, you will need to enter a simple program for a one-independent variable chi-square test. Enter the program found in Appendix A, *Using Technology,* for a one-way chi-square calculation into your calculator. When prompted by the program, enter the number of categories, and for each category, enter the observed and expected values. When the calculated chi-square value is displayed, use Table 11.8 *Chi-Square Sampling Distribution,* to determine the probability that differences in frequencies were due to chance.

*When you have **two** independent variables:*

1. In the **MATRIX** mode of your calculator, create the size of the matrix needed for your data, for example, **2 x 3** for a study of the effect of gender (two, male and female) and preference for certain TV shows (three shows). (See Appendix A, *Using Technology*, for additional help in using the graphing calculator).
2. Enter the *observed* data in each cell of the matrix; the calculator will calculate the *expected* values. In some brands of calculators, you must create a second matrix for the *expected* values.
3. In the **STAT** mode of your calculator, select **TESTS** and choose chi-square.
4. Enter the letter name for the matrix or matrices, and then calculate.
5. The calculated chi-square value, the probability the differences in frequencies are due to chance, and the degrees of freedom will be displayed.

Designing Complex Experiments

© 1998 PhotoDisc, Inc.

Objectives

- Diagram complex experimental designs including repeated measures over time or subjects, multiple independent variables, and correlation of variables.
- Describe positive features of complex experimental designs and select the most appropriate design for a given experiment.
- Construct appropriate data tables and graphs for data derived from complex experimental designs.
- Write appropriate paragraphs of results and conclusions for investigations involving complex experimental designs.
- Assist individual students with complex research questions to select an appropriate experimental design.
- Assist individual students in preparing appropriate data tables, graphs, results, and conclusions for complex experimental designs.

National Standards Connections

- Design and conduct scientific investigations (NSES).
- Design a statistical experiment to study a problem, conduct the experiment, and interpret and communicate the outcomes (NCTM).
- Use curve fitting to predict from data (NCTM).
- Transform data to aid in data interpretation and prediction (NCTM).

For the beginning researcher, experiments with one independent variable are recommended. Basic concepts of experimental design and ways to improve these designs are described in Chapters 1 and 2. With experience, students can handle more complex designs involving repeated measures over time or subjects, multiple variables, and the correlation of variables. For each of these designs, a scenario, an experimental design dia-gram, and graphical procedures for analyzing data are presented in this chapter.

REPEATED MEASURES OVER TIME

One simple way to modify and improve an experiment with one independent variable is to obtain multiple measures of the dependent variable over time. An experiment to determine the influence of earthworms on soil quality would be enhanced by reporting results weekly, rather than at the end of a two-month period. Similarly, the influence of aerobic exercise on resting pulse rate can be more accurately assessed with monthly measurements, rather than one at the end of a year. This design is particularly effective when differential effects over time are hypothesized. For example, one fertilizer may act more quickly than another to promote plant growth, yet they may produce equivalent growth at six weeks.

Scenario

Ronald raised chickens as a hobby. In his science class, he studied the importance of a balanced diet. The feed Ronald normally purchased contained a greater percentage of carbohydrates and fats and a lower percentage of protein than recommended. Ronald decided to mix his own feed to include the recommended daily allowances. He hypothesized that the chicks fed a balanced diet would have a smaller mass and be more active. Thirty-five chicks were fed the nonnutritionally balanced feed; an additional 35 chicks were fed the home-mixed nutritionally balanced feed. The chicks were the same age, received the same amount of food and water, and were reared under

identical climate and space conditions. The mass and activity level of the chicks were measured weekly. Activity level was rated on a scale of 1 to 3, with 1 = inactive, 2 = moderately active, and 3 = very active. To ensure no adverse effects on the chicks, a local agricultural extension agent served as Ronald's mentor. Remember, all projects involving vertebrates must be approved in advance and supervised by a mentor; the experiment must be discontinued immediately if adverse effects occur.

Design and Analysis

Use a modified experimental design diagram format, as illustrated in Table 12.1 *Experimental Design Diagram* to communicate both the independent variable and the time intervals. Depict time across the top of the rectangle and the levels of the independent variable along the side. Although these parameters can be switched, the data display will be facilitated by depicting time horizontally.

Transform the experimental design diagram into a data table by adding a column for descriptive information that includes the appropriate measures of central tendency, variation, and number. Data tables for chick mass (quantitative) and activity rate (qualitative) are illustrated in Table 12.2 *Data Display: Diet and Chickens*. Because

TABLE 12.1　Experimental Design Diagram

Title: The Effect of Diet on Chicken Growth and Behavior
Hypothesis: If chicks are fed a balanced diet, then they will have a smaller mass and be more active.

Type of diet	Time (weeks)					
	1	2	3	4	5	6
Non-balanced (Control)	35 chicks	35 chicks	35 chicks	35 chicks	35 chicks	35 chicks
Balanced (Experimental)	35 chicks	35 chicks	35 chicks	35 chicks	35 chicks	35 chicks

DV:　Mass (g)
　　　Activity (scale of 1 to 3)
C:　Age
　　　Food
　　　Water
　　　Climate
　　　Space

TABLE 12.2   Data Display: Diet and Chickens

Diet & Mass
Data Table and Graph A—Effect of Diet on Chick Mass (g)

Type of diet	Descriptive information	Time (weeks) 1	2	3	4	5	6
Non-balanced diet	Mean Range Number	610 16 35	648 20 35	698 19 35	750 22 35	820 30 35	880 33 35
Balanced diet	Mean Range Number	612 18 35	630 19 35	660 21 35	680 18 35	720 21 35	745 19 35

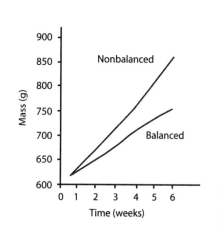

Diet & Activity
Data Table & Graph B—Effect of Diet on Chick Activity

Type of diet	Descriptive information	Time (weeks) 1	2	3	4	5	6
Nonbalanced diet	Median	3	3	2	2	1	1
	Frequency Distribution						
	Activity 3	20	25	10	8	5	6
	Activity 2	10	7	18	12	12	7
	Activity 1	5	3	7	15	18	22
	Number	35	35	35	35	35	35
Balanced diet	Median	3	3	2	2	2	2
	Frequency Distribution						
	Activity 3	20	23	17	14	11	10
	Activity 2	8	6	13	18	20	22
	Activity 1	4	6	5	3	4	3
	Number	35	35	35	35	35	35

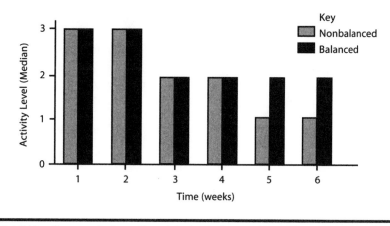

© 2000 by Kendall/Hunt Publishing Company, Cothron, Giese, & Rezba, *Students and Research*.

repeated measures are involved, the variables displayed on the axes of the graph must be altered. Display time intervals on the X axis and mass as the dependent variable on the Y axis. Using different symbols, graph the mass of the chicks fed the nonbalanced feed (control) and then graph the mass of the chicks fed the balanced feed. Similarly, construct a bar graph to display the activity level of chicks fed the two diets. Use procedures described in Chapter 9, *Communicating Descriptive Statistics,* for writing results and conclusions. For statistical analysis, consult reference texts on the correlated *t* test or one-way analysis of variance with repeated measures.

Conducting and Analyzing an Experiment

Read Investigation 12.1 *Holding the Heat* (on page 189). Design a class experiment to test the effectiveness of three different insulators. Draw an experimental design diagram. Conduct the experiment and summarize the data using appropriate tables, graphs, and paragraphs.

REPEATED TREATMENTS OVER SUBJECTS

By exposing the same subjects to different treatments, the experimenter can minimize experimental error resulting from variations within subjects. Repeated treatment designs are particularly effective in psychological and biological studies involving higher organisms. For example, the most effective time for learning could be investigated by determining 60 subjects' rates of learning nonsense syllables at three different times during the day. With this design, each student serves as his or her own control. Thus, genetic and environmental factors are minimized. Because nonliving matter exhibits less variation, repeated treatment designs are less common in the physical and earth sciences.

Scenario

Although Sarah's classmates learned the advantages of studying in a quiet place, the majority of the students continued to complete homework while watching television or listening to music. Sarah hypothesized that ability to solve math-

ematics problems would decrease with an increase in the number of stimuli in the environment. Sarah taped three 30-minute segments of television programs that were equivalent in conversation, music, and screen action. She developed four equivalent mathematics tests (20 items) on decimals and percents. Fifteen students were randomly selected from the class; parental permission to participate was secured. Students completed a mathematics test while exposed to no stimuli, to an auditory tape, to a video, and to a tape of the television programs. The form of the mathematics test, type of stimuli, and order of presentation of the stimuli were randomized. Time for completion was recorded. Tests were administered at the same time of day in identical test sites. The sound level of the auditory stimuli and the screen size of the visual stimuli remained constant.

Design and Analysis

As diagrammed in Table 12.3 *Experimental Design Diagram,* place the levels of the independent variable across the top of the experimental design diagram and the subjects along the side. List the dependent variables and constants.

Calculate the descriptive statistics for each treatment group, including measures of central tendency (mean) and variation (standard deviation or range). Depict the results graphically. As illustrated in Table 12.4 *The Effect of Stimuli on Time (sec) to Solve Mathematics Problems,* individual subjects' measures are reflected in the raw data but not in the summary data table or graph. Repeated measures on subjects are an important control feature of the experimental design and are emphasized in the methods and materials section, rather than the results section of the research report. For statistical analysis, consult reference texts on the correlated *t* test or one-way analysis of variance with repeated measures.

Conducting and Analyzing an Experiment

Read Investigation 12.2 *Practice Makes Perfect* (on page 191). Design a class experiment to test the impact of practice on the ability to perform a task. Draw an experimental design diagram. Conduct the experiment and summarize the data using appropriate tables, graphs, and paragraphs.

TABLE 12.3   Experimental Design Diagram

Title: The Effect of Stimuli on Mathematics Problem Solving
Hypothesis: Time for completion of mathematics tests will increase as the number of stimuli in the environment increases.

Subjects	Type of stimuli			
	None (Control)	Auditory	TV Tape	Video
1				
2				
3				
•				
•				
•				
15				

DV: Time to complete (sec)
C: Length and difficulty of test
   Length, complexity, and intensity of stimuli
   Time and place of administration

TWO INDEPENDENT VARIABLES

When the literature review suggests that two factors may interact to influence the outcome, an experimental design involving two independent variables is appropriate. For example, concentration and time of application may influence the effectiveness of a weed killer. In experiments with two independent variables, both main and interaction effects are hypothesized. **Main effects** refer to the action of each independent variable alone, whereas **interaction** refers to the combined action of the variables. Experiments with two independent variables are not appropriate unless an interaction is hypothesized.

- **Main Effect:** Higher concentrations of Herbicide X will result in a greater death rate of weeds.
- **Main Effect:** Early morning application of Herbicide X will result in a greater death rate of weeds.
- **Interaction:** Time of application and concentration will interact to affect the death rate of weeds, for example, lower concentrations of Herbicide X will effectively kill weeds in the morning while higher concentrations will be required in the afternoon.

Scenario

Larry read several articles about the effects of playing video games. One article indicated that playing video games increased an individual's stress level. A second article suggested that older people displayed more stress than younger people. The third article indicated that stress level was a function of the type of video game and the relevance of the game to real-life situations. Larry decided to investigate the effect of both age and type of video game on stress level as measured by the pulse rate of the individual. Larry randomly selected 45 subjects (SS) subdivided into three age groups: (a) 10–12 years (15 SS), (b) 15–17 years (15 SS), and (c) 20–23 years (15 SS). Five subjects within each age group were then assigned to play a video game: (a) Wac-Man (5 SS), (b) Busdec (5 SS), and (c) Cosmos (5 SS). Each individual played the game in the same setting for 20 minutes. Larry hypothesized that individuals would place greater importance on winning Wac-Man, a popular recreation game. Successively less impor-

TABLE 12.4   The Effect of Stimuli on Time (sec) to Solve Mathematics Problems

Data table for raw data:

Subjects	Type of stimuli			
	None	Auditory	TV Tape	Video
1	20.0	35.0	21.0	36.0
2	17.0	20.0	16.0	22.0
3	22.0	27.0	24.0	30.0
4	14.0	18.0	15.0	18.0
5	13.0	19.0	14.0	18.0
6	25.0	30.0	25.0	29.0
7	18.0	28.0	22.0	30.0
8	12.0	15.0	11.0	14.0
9	15.0	20.0	19.0	20.0
10	16.0	21.0	19.0	24.0
11	21.0	30.0	23.0	29.0
12	17.0	23.0	19.0	24.0
13	18.0	20.0	19.0	21.0
14	20.0	28.0	22.0	29.0
15	15.0	24.0	19.0	28.0

Data table for summarized data:

Descriptive information	Type of stimuli			
	None	Auditory	TV Tape	Video
Mean	17.5	23.9	19.2	24.8
Range	13.0	20.0	14.0	22.0
Maximum	25.0	35.0	25.0	36.0
Minimum	12.0	15.0	11.0	14.0
Number	15	15	15	15

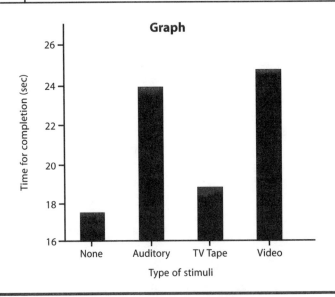

Graph

© 2000 by Kendall/Hunt Publishing Company, Cothron, Giese, & Rezba, *Students and Research*.

tance would be placed upon winning Busdec, a business simulation, and Cosmos, a space game. He also hypothesized that age and type of video game would interact with older students exhibiting greater stress with Busdec. Larry stated three research hypotheses to guide the experimental design and analysis:

- **Main Effect:** The age of the subject will influence pulse rate after playing a video game with older subjects exhibiting the highest pulse rate.
- **Main Effect:** The type of video game will influence the pulse rate of subjects, such as Wac-Man > Busdec > Cosmos.
- **Interaction:** The subject's age and type of video game will interact to influence the subject's pulse rate. For example, younger subjects will exhibit the highest pulse rate with Wac-Man, and older subjects with Busdec. Reaction to Cosmos will be equivalent across age groups.

Design and Analysis

Modify the experimental design diagram to show one independent variable across the top and the second along the side (see Table 12.5 *Experimen-*

tal Design Diagram). Generally, the treatment variable is shown across the top and a categorical variable along the side, such as age, intelligence, gender. Remember to include the dependent variables, constants, and number of repeated trials in the experimental design.

To determine the main and interactive effects of the two variables, measures of central tendency and variation must be computed for each cell, each horizontal row, and each vertical column of the raw data table. Construct a reduced data table from the raw data table as described next and illustrated in Table 12.6 *The Effect of Age and Video Game on Pulse Rate*. For statistical analysis, consult a reference text on two-way analysis of variance.

- **Main Effect for Type of Video Game:** Treat the data as if they were from an experiment with one independent variable (type of video game) by calculating the mean and standard deviation for each vertical column. Place the summary statistics under the appropriate column.
- **Main Effect for Influence of Age:** Treat the data as if they were from an experiment with one independent variable (age) by calculating the mean and standard deviation for each

TABLE 12.5 **Experimental Design Diagram**

Title: The Effect of Age and Type of Video Game on Pulse Rate
Hypothesis:
- Older subjects will exhibit higher pulse rates.
- Type of video game will affect pulse rate.
- Subject's age and type of video game will interact to affect pulse rate.

IV: Age group	IV: Type of Video Game		
	Wac-Man	**Cosmos**	**Busdec**
10–12 years	5 people	5 people	5 people
15–17 years	5 people	5 people	5 people
20–23 years	5 people	5 people	5 people

DV: Change in pulse rate
C: Time of play (20 min)
Setting

TABLE 12.6   The Effect of Age and Video Game on Pulse Rate

Data table for raw data:

Age group	Wac-Man	Busdec	Cosmos
	Type of video game		
10–12 years	80.0 90.0 82.0 85.0 78.0	82.0 88.0 84.0 87.0 80.0	90.0 98.0 100.0 88.0 85.0
15–17 years	78.0 94.0 84.0 81.0 80.0	88.0 95.0 80.0 82.0 85.0	80.0 88.0 76.0 83.0 80.0
20–23 years	88.0 90.0 87.0 95.0 94.0	78.0 75.0 70.0 72.0 73.0	60.0 62.0 64.0 68.0 62.0

Data table for summarized data:

Age group	Wac-Man	Busdec	Cosmos	
	Type of video game			
10–12 years	$\bar{X} = 83.0$ SD = 4.2 n = 5	$\bar{X} = 84.2$ SD = 2.9 n = 5	$\bar{X} = 92.2$ SD = 5.8 n = 5	$\bar{X} = 86.5$ SD = 6.1 n = 15
15–17 years	$\bar{X} = 83.4$ SD = 5.6 n = 5	$\bar{X} = 86.0$ SD = 5.3 n = 5	$\bar{X} = 81.4$ SD = 4.0 n = 5	$\bar{X} = 83.6$ SD = 5.3 n = 15
20–23 years	$\bar{X} = 90.8$ SD = 3.2 n = 5	$\bar{X} = 73.6$ SD = 2.7 n = 5	$\bar{X} = 63.2$ SD = 2.7 n = 5	$\bar{X} = 75.9$ SD = 11.7 n = 15
	$\bar{X} = 85.7$ SD = 5.7 n = 15	$\bar{X} = 81.3$ SD = 6.7 n = 15	$\bar{X} = 78.9$ SD = 13.3 n = 15	$\bar{X} = 81.9$

horizontal row. Record the summary statistics for each row in the right-hand margin of the table.

- **Interactive Effect of Type of Video Game and Age:** Compute the mean and standard deviation for each cell within the experimental design.

Tables and Graphs

Methods for constructing data tables and graphs to display main effects are described in Chapters 5 and 8. The interaction of the variables is also determined by graphic analysis. One independent variable (for example, type of video game) is depicted along the X axis and the dependent variable on the Y axis. The mean pulse rate for individuals, aged 10–12 years, is graphed. Using different symbols, data for the other two age groups are displayed. No interaction exists if the patterns of lines on the graph are the same. An interaction exists when graphed lines cross or diverge from parallel alignment. Similarly, an interaction exists when sets of bars on a graph have different patterns. A bar graph is the appropriate graph for Larry's data because type of video game is categorical data and the intervals between the levels of the independent variable are meaningless (see Chapter 5). In Larry's experiment, an interaction occurred between age and type of video game. Pulse rate of ages 10–12 was less when playing Wac-Man, while pulse rate of ages 20–23 was greater. The pulse rate pattern for ages 20–23 is opposite the observed pulse rates for the other two age groups. (See Table 12.7 *Data Display, Age and Video Games on Pulse Rate* for illustrative data tables and graphs). When writing the results, the findings for each of the main effects and the interaction must be described.

Results

The effect of type of video game on pulse rate is summarized in Table and Graph 12.7A. Pulse rate was highest in Wac-Man players (85.7), followed by Busdec (81.3), and Cosmos (78.9) players. Variance in pulse rate was greater among Cosmos players. The hypothesized influence of type of video game on pulse rate was supported.

The influence of age on pulse rate is summarized in Table and Graph 12.7B. Substantially

lower pulse rates occurred in ages 20–23 (75.9) than among individuals aged 15–17 years (83.6) or 10–12 years (86.5). Variation in pulse rates was greatest among ages 20–23. The data contradicted the research hypothesis that older subjects would exhibit progressively higher pulse rates.

Interactive effects of the subject's age and type of video game are summarized in Table and Graph 12.7C. Higher pulse rates occurred among older subjects (20–23 years) playing Wac-Man and younger subjects (10–12; 15–17 years) playing Cosmos. The experimental data do not support the hypothesized pulse increase with older subjects playing Busdec or younger subjects playing Wac-Man. Substantial differences were found among the pulse rates of different aged subjects playing Cosmos. The data do not support the hypothesis that Cosmos players would have equivalent pulse rates. Instead, substantial interaction occurred with subjects aged 20–23 exhibiting the highest pulse rate with Wac-Man, ages 15–17 the highest pulse rate with Busdec, and ages 10–12 the highest pulse rate with Cosmos. In general, these findings support the hypothesis of an interactive effect of the subject's age and type of video game on pulse rate, but not the specifics of the interaction.

THREE INDEPENDENT VARIABLES

Experiments involving three independent variables are appropriate when three factors may potentially interact to influence the value of the dependent variable. Respiratory rate in fish, for example, may be affected more strongly by the combination of thermal, phosphate, and acid rain pollution than would be predicted by the action of each variable alone. When an experiment includes three independent variables, then three main effects, three double interactions, and one triple interaction can be investigated. Experimental designs with three independent variables are appropriate only when a triple interaction is hypothesized.

Main Effects — Thermal Pollution
Phosphate Pollution
Acid Rain Pollution

TABLE 12.7   Data Display, Age and Video Games on Pulse Rate

<table>
<tr><td colspan="4" align="center">A
Effect of Type of Video Game on Pulse Rate
(beats/minute)</td></tr>
<tr><td rowspan="2">Descriptive
information</td><td colspan="3" align="center">Type of video game</td></tr>
<tr><td>Wac-Man</td><td>Busdec</td><td>Cosmos</td></tr>
<tr><td>Mean</td><td>85.7</td><td>81.3</td><td>78.9</td></tr>
<tr><td>Standard
deviation</td><td>5.7</td><td>6.7</td><td>13.3</td></tr>
<tr><td>Number</td><td>15</td><td>15</td><td>15</td></tr>
</table>

<table>
<tr><td colspan="4" align="center">B
Effect of Age on Pulse Rate
(beats/minute)</td></tr>
<tr><td rowspan="2">Descriptive
information</td><td colspan="3" align="center">Age Group</td></tr>
<tr><td>10–12
yrs.</td><td>15–17
yrs.</td><td>20–23
yrs.</td></tr>
<tr><td>Mean</td><td>86.5</td><td>83.6</td><td>75.9</td></tr>
<tr><td>Standard
deviation</td><td>6.1</td><td>5.3</td><td>11.7</td></tr>
<tr><td>Number</td><td>15</td><td>15</td><td>15</td></tr>
</table>

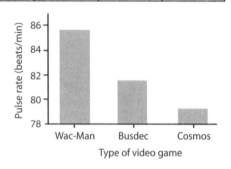

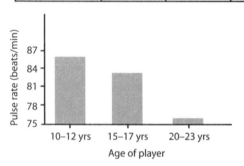

<table>
<tr><td colspan="4" align="center">C
Effect of Age and Type of Video Game on Pulse Rate
(beats/minute)</td></tr>
<tr><td rowspan="2">Age group</td><td colspan="3" align="center">Type of video game</td></tr>
<tr><td>Wac-Man</td><td>Busdec</td><td>Cosmos</td></tr>
<tr><td>10–12 years</td><td>$\overline{X} = 83.0$
SD = 4.2</td><td>$\overline{X} = 84.2$
SD = 2.9</td><td>$\overline{X} = 92.2$
SD = 5.8</td></tr>
<tr><td>15–17 years</td><td>$\overline{X} = 83.4$
SD = 5.6</td><td>$\overline{X} = 86.0$
SD = 5.3</td><td>$\overline{X} = 81.4$
SD = 4.0</td></tr>
<tr><td>20–23 years</td><td>$\overline{X} = 90.8$
SD = 3.2</td><td>$\overline{X} = 73.6$
SD = 2.7</td><td>$\overline{X} = 63.2$
SD = 2.7</td></tr>
</table>

n = 5 subjects/cell

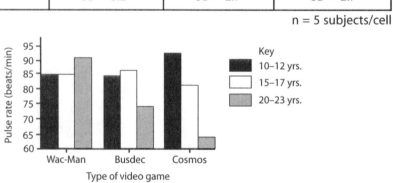

© 2000 by Kendall/Hunt Publishing Company, Cothron, Giese, & Rezba, *Students and Research*.

Double Interactions	Thermal and Phosphate Pollution
	Thermal and Acid Rain Pollution
	Phosphate and Acid Rain Pollution
Triple Interaction	Thermal, Phosphate, and Acid Rain Pollution

Scenario

Nina was interested in the effectiveness of aerosol and nonaerosol forms of Superwash and Out! prewash cleaners in removing common stains, such as catsup, mustard, grass, new motor oil, soot, and used motor oil. She hypothesized that

- **Main Effect:** Plant stains (catsup, mustard) will be easier to remove than hydrocarbon stains (new motor oil, soot, used motor oil).
- **Main Effect:** Superwash will remove stains more effectively than Out! or the detergent alone (control).
- **Main Effect:** Aerosol prewashes will remove stains more effectively than nonaerosol prewashes.
- **Interaction:** Nonaerosol types of Superwash will be more effective on plant stains, aerosol forms of Out! will be more effective on hydrocarbon stains.

Cotton-polyester sheeting was cut into 15 cm × 15 cm squares and a ring drawn in the center. One scoop of each stain was applied, evenly spread over the circle, and allowed to dry for 24 hours. Nine samples of each stain were treated with aerosol Out!, nonaerosol Out!, aerosol Superwash, and nonaerosol Superwash. Nine samples of each stain remained untreated and served as a control. All samples were washed on the regular cycle with warm water. The recommended amount of Brand X detergent was used. After drying for 24 hours, the degree of stain removal was rated on a 5-point scale, with a score of 5 representing total removal and 1 representing no removal.

Design and Analysis

In Table 12.8 *Experimental Design Diagram and Raw Data,* a three-way experimental design is illustrated; stain removal ratings for the nine samples/cell are included to facilitate data analysis in succeeding sections. Depict one independent variable (type of stain) across the top of the rectangle and the remaining two variables (type and brand of prewash) along the side. The number of trials is noted within each cell.

In the experiment, Nina used an ordered categorical scale (1–5) to rate cleaning effectiveness. Thus, the appropriate measure of central tendency is the median; variation in data is expressed through a frequency distribution. Begin compiling data by computing a frequency distribution and median for each cell. Then, compute the frequency distribution and median value for the following:

- All samples washed within a stain category, e.g., mustard, catsup . . . used oil (*Place the value under the appropriate vertical column*).
- All samples washed with aerosol and all samples washed with nonaerosol prewash (*Place the value in the right margin of the table*).
- All samples washed with Superwash and all samples washed with Out! (*Place the value in the far right margin of the table*).
- All samples with no prewash—the control. (*Place the value in the right margin of the table*).

A reduced data table with median values for each cell and for each independent variable is provided in Table 12.9 *Data Analysis, Prewashes and Stain Removal (Scale of 1 to 5).* For statistical analysis, consult a reference text on three-way analysis of variance.

Tables and Graphs

For each of the independent variables, construct a data table and graph. From the data displays (see Table 12.10 *The Effect of Types of Stains, Prewash Brands, and Prewash Types on Stain Removal (Scale of 1 to 5),* determine the effect of the three variables acting alone.

TABLE 12.8   Experimental Design Diagram and Raw Data

The Effect of Types of Stains, Prewash Brands, and Prewash Types on Stain Removal (Scale of 1 to 5)

● Plant stains (catsup, mustard) will be easier to remove than hydrocarbon stains (new motor oil, soot, used motor oil).

● **Superwash** will remove stains more effectively than **Out!** or the detergent alone (control).

● Nonaerosol types of **Superwash** will be more effective on plant stains, while aerosol forms of **Out!** will be more effective on hydrocarbon stains.

IV: Type of stain (mustard, catsup, grass, new oil, soot, used oil)
Brand of prewash (**Superwash, Out!**, none)
Type of prewash (aerosol, nonaerosol, none)

Prewashes		Stain					
Types	**Brands**	**Mustard**	**Catsup**	**Grass**	**New Motor Oil**	**Soot**	**Used Motor Oil**
Aerosol	Superwash	4 4 4 5 3 5 5 5 4	4 5 4 5 5 5 4 5 4	4 3 5 5 4 4 3 4 4	2 3 2 3 3 3 2 3 3	1 2 2 2 2 3 3 3 3	2 1 3 2 2 2 2 3 2
Aerosol	Out!	5 5 4 4 4 4 4 4 5	3 5 4 4 4 3 3 4 5	3 3 3 2 4 2 3 2 3	4 3 2 4 4 3 3 4 4	3 3 3 2 4 2 3 3 3	3 4 3 4 3 4 4 4 4
Nonaerosol	Superwash	5 4 4 5 3 5 5 5 5	5 5 4 4 5 5 4 5 5	4 4 4 3 2 3 3 3 3	2 3 2 3 2 2 2 3 3	2 1 3 2 3 2 2 2 2	2 3 2 2 2 3 3 2 1
Nonaerosol	Out!	4 5 5 5 4 4 5 4 4	5 3 4 4 4 3 4 5 4	2 3 3 2 1 2 2 2 2	2 3 4 2 3 2 3 3 2	3 2 4 4 3 3 3 3 3	3 3 2 2 3 3 2 2 3
None	None (control)	4 2 4 4 3 3 3 3 3	3 4 3 3 4 3 4 3 4	2 3 3 2 3 3 3 2 3	2 1 2 1 1 1 1 2 1	1 3 1 1 2 2 2 1 1	2 1 1 3 1 1 1 2 2

DV: Amount of stain removed (Scale of 1 to 5, 5 = total removal and 1 = no removal)

C: Duration of wash
Temperature of water
Wash cycle
Amount of stain
Fabric stained
Drying time

© 2000 by Kendall/Hunt Publishing Company, Cothron, Giese, & Rezba, *Students and Research.*

TABLE 12.9   Data Analysis, Prewashes and Stain Removal (Scale of 1 to 5)

Prewashes		Stain						
Types	Brands	Mustard	Catsup	Grass	New Motor Oil	Soot	Used Motor Oil	
Aerosol	Superwash	4	5	4	3	2	2	M = 3
Aerosol	Out!	4	4	3	4	3	4	M = 3
Non-aerosol	Superwash	5	5	3	2	2	2	M = 3
Non-aerosol	Out!	4	4	2	3	3	3	M = 3
None (Control)	None (Control)	3	3	3	1	1	1	$M_s = 3$ $M_o = 3$ $M = 2$
		M = 4	M = 4	M = 3	M = 3	M = 2	M = 2	

- Effect of Type of Stain on Cleaning Effectiveness
- Effect of Brand of the Prewash on Cleaning Effectiveness
- Effect of Type of Prewash on Cleaning Effectiveness

To determine the interaction of variables, depict one independent variable (for example, type of stain) on the X axis and the dependent variable (such as cleaning effectiveness) on the Y axis. Using two different sets of symbols, graph the cleaning effectiveness of aerosol Superwash and Out! Repeat the process for the nonaerosol sprays and the control (see Table 12.10). Describe the effect of each variable acting alone and in combination. Relate the findings to the research hypotheses.

Results

The influence of type of stain on cleaning effectiveness is illustrated in Table and Graph 12.10A. The highest median cleaning (4) occurred with mustard and catsup stains, followed by grass/new oil (median = 3), and soot/used oil (median = 2). In general, the research hypothesis that plant stains would be more effectively cleaned than hydrocarbon stains was supported. The exception was grass, which equaled new oil in cleaning difficulty. As depicted in Table and Graph 12.10B, both brands of prewash were more effective than the control. Equivalent cleaning occurred with Superwash and Out! The hypothesized superiority of Superwash was not supported by the data. Aerosol and nonaerosol cleaners were equally effective in removing stains (see Table 12.10C). The data did not support the hypothesized superiority of aerosol prewashes. The type and brand of prewash influenced cleaning effectiveness. Aerosol and nonaerosol Superwash were more effective with plant stains and aerosol and nonaerosol Out! with hydrocarbons. The following lists indicate the most effective cleaner for each type of stain:

Stain	Most Effective Cleaners
Mustard	Nonaerosol Superwash
Catsup	Nonaerosol or aerosol Superwash
Grass	Nonaerosol Superwash
New oil	Aerosol Out!
Soot	Aerosol or nonaerosol Out!
Used oil	Aerosol Out!

TABLE 12.10   The Effect of Types of Stains, Prewash Brands, and Prewash Types on Stain Removal (Scale of 1 to 5)

The Effect of Types of Stains, Prewash Brands, and Prewash Types on Stain Removal (Scale of 1 to 5)

A
Main Effect, Type of Stain

Descriptive information	Type of stain					
	Mustard	Catsup	Grass	New oil	Soot	Used oil
Median	4	4	3	3	2	2
Frequency distribution						
5	17	15	2	0	0	0
4	20	20	9	6	3	6
3	7	10	20	17	19	14
2	1	0	13	16	16	18
1	0	0	1	6	7	7
Number	45	45	45	45	45	45

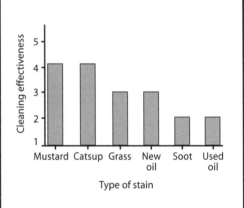

B
Main Effect, Brand of Prewash

Descriptive information	Brand of prewash		
	Superwash	Out!	None (Control)
Median	3	3	2
Frequency distribution			
5	23	11	0
4	21	36	7
3	30	39	17
2	30	21	14
1	4	1	16
Number	108	108	54

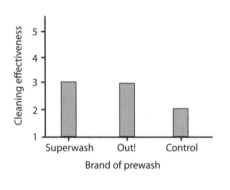

C
Main Effect, Type of Prewash

Descriptive information	Type of prewash		
	Aerosol	Nonaerosol	None (Control)
Median	3	3	2
Frequency distribution			
5	16	18	0
4	36	21	7
3	35	35	18
2	19	31	12
1	2	3	16
Number	108	108	54

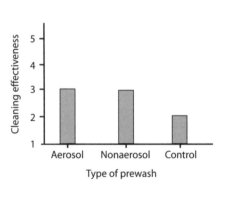

(continued on the following page)

TABLE 12.10 (continued)

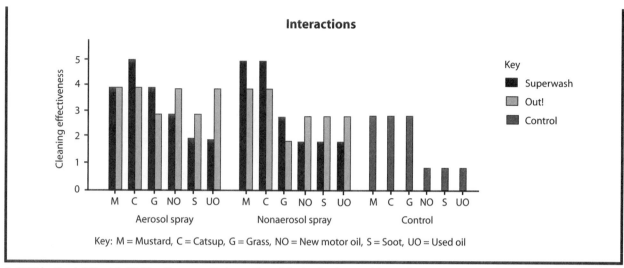

© 2000 by Kendall/Hunt Publishing Company, Cothron, Giese, & Rezba, *Students and Research.*

When both the type and brand of cleaner were considered, nonaerosol Superwash proved superior with plant stains and aerosol Out! with hydrocarbon stains. These findings supported the research hypothesis of differential effectiveness of type and brand of prewash with various stains.

Conclusion

In general, plant stains proved easier to clean than hydrocarbon stains. Based upon chemical structure, these findings were predicted. Plants consist of complex carbohydrates containing oxygen, carbon, and hydrogen atoms; charged areas occur along the molecule. Hydrocarbons consist of long chains of hydrogen and carbon, which tend to repel both water and detergents; the molecule is nonpolar and does not have charged sections. The charged areas in carbohydrates promote bonding among the stain, detergent, and water, and thereby makes stain removal easier. Prewash treatments proved more effective than the detergent alone (control). Both types and brands were equivalent in general stain removal. Nonaerosol Superwash proved most effective with plant stains, and aerosol Out! with hydrocarbon stains. Prewashes act by promoting the breakdown of the stain and increasing the bonding of the detergent and water with the stain. Because substances with similar bonding are more soluble, Superwash must contain more charged (polar) particles while Out! contains more noncharged particles. The foaming agent may also increase the solubility of hydrocarbon stains, or greater wetting with nonaerosol sprays may promote removal of plant stains. The experiment could be improved by including more brands and by increasing the sample size. Additional studies should focus on stains that did not readily fit the pattern, for example, grass and new motor oil.

Conducting and Analyzing an Experiment

Read Investigation 12.3 *Fruit Appeal* (on page 193). Design a class experiment to determine the impact of various factors on the oxidation of fruit. Draw an experimental design diagram. Conduct the experiment and summarize the data using appropriate tables, graphs, and paragraphs.

CORRELATION

Researchers are frequently concerned with predicting the value of one variable (Y) from knowledge of a second variable (X). For example, college admissions officers predict success in college from SAT scores, and astronomers predict disturbances in radio transmission from the number of sunspots. Today, climatologists seek to predict the warming of the Earth's atmosphere from hydro-carbon levels. Prediction of one variable from another variable is possible only when a correlation exists.

If two variables, X and Y, are not related, a zero correlation exists. Knowledge of X tells the researcher nothing about Y, and vice versa. When the values of the two variables are graphed, as illustrated in Figure 12.1, no systematic pattern results. Systematic patterns emerge when a nonzero correlation exists. If the value of one variable (Y) increases with the value of the second variable (X), a positive correlation exists. The more closely data points approach a linear pattern, the stronger the correlation. When the value of one variable decreases (Y) as the value of the second variable increases (X), a negative correlation exists. The stronger the correlation between the variables, the more accurately one variable can be predicted from knowledge of the other (see Figure 12.1). Mathematical techniques, as well as graphic representations also exist for expressing the correlation between variables. Computed values range from –1 to +1. If the correlation between variables is either +1 or –1, perfect prediction is possible. For statistical analysis of data, consult a reference text for "Pearson's Product Moment Correlation" or "Spearman Rank Order Correlation."

Scenario

Rita Lynn Gilman determined the wood production of loblolly pine trees over a 40-year interval. A major interstate was constructed adjacent to the forest during the third and fourth decade. Rita Lynn had previously determined that wood production in that forest was significantly less than in a comparable forest fronted by rural roads. Because wood production substantially declined after construction of the interstate, Rita Lynn wondered whether wood production could be predicted from traffic counts.

Design and Analysis

In correlation studies, the value of each variable (X and Y) is simultaneously determined for each subject, year, and so on. For example, the consumption of saturated fats and blood cholesterol level could be determined for each of 5,000 sub-

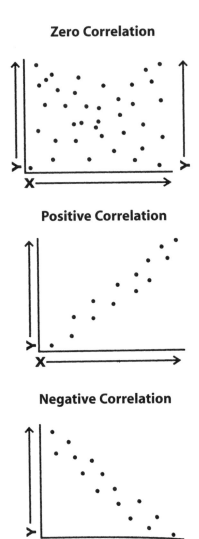

Figure 12.1 Scatterplot.

jects or the number of sunspots and intensity of northern lights could be secured from 50 years of records. The experimental design diagram depicts the framework for the experiment (year, subject, and so on) and the specific variables measured. One variable (X) is designated as the independent (predictor) variable and the second as the dependent variable (Y). In Rita Lynn's experiment, the experimental framework was the 18-year interval. For each year, traffic count acted as the independent (predictor) variable and wood production as the dependent variable. Table 12.11 *Raw Data on Traffic Count and Tree Ring Width* contains the raw data.

Determine an appropriate scale for depicting the independent variable (for example, traffic

TABLE 12.11   Raw Data on Traffic Count and Tree Ring Width

Year	Independent Variable — Traffic Count (Vehicles/year)	Dependent Variable — Wood Production Tree Ring Area (mm²)
1965	26,240	1149.42
1966	26,010	1011.50
1967	27,325	966.680
1968	30,010	819.769
1969	30,090	931.373
1970	28,795	824.337
1971	30,740	864.514
1972	32,930	1021.28
1973	33,940	902.085
1974	31,690	866.515
1975	34,040	932.353
1976	35,540	869.129
1977	38,440	649.712
1978	41,760	605.717
1979	43,150	506.946
1980	41,070	560.024
1981	42,330	617.351
1982	43,700	437.180
1983	44,140	348.047

count) on the X axis and the dependent variable (for example, wood production) on the Y axis. Plot each data point on the graph. Examine the scatterplot to determine whether a zero, positive, or negative correlation exists. In this instance, a negative correlation existed (see Figure 12.2), because wood production decreased as traffic count increased.

Conducting and Analyzing an Experiment

Read Investigation 12.4 *Collapsing Bridges* (on page 194). Design a class experiment to investigate the impact of bridge thickness on collapsing point. Draw an experimental design diagram. Conduct the experiment and summarize the data using appropriate tables, graphs, and paragraphs.

REFERENCES

Gilman, R. L. *The possible effects of pollutants on the wood production of Loblolly pine trees.* Paper presented at the meeting of the Virginia Junior Academy of Science, First Place, Environmental Science A & American Academy of Science Presentor, Harrisonburg, VA: James Madison University.

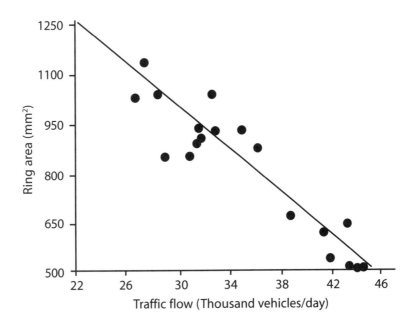

Figure 12.2 Relationship of Traffic Flow to Yearly Wood Production.

Glenberg, A. (1996). *Learning from data: An introduction to statistical reasoning* (2nd ed.). Mahwah, NJ: Lawrence Erlbaum Associates.

Landwehr, J.M. & Watkins, A.E. (1994). *Exploring data*. A component of the Quantitative Literacy Series. Palo Alto, CA: Dale Seymour Publications.

Landwehr, J.M., Swift, J., & Watkins, A.E. (1998). *Exploring surveys and information from samples.* A component of the Data-Driven Mathematics Series. Palo Alto, CA: Dale Seymour Publications.

Lyon, C. G. *The synergistic effects of acid rain, thermal pollution, and phosphate pollution on common shiner minnows (Notropis cornutus).* Paper presented at the meeting of the Virginia Junior Academy of Science, First Place, Environmental Science B and R. J. Rowlett Award for Best Research Paper, Harrisonburg, VA: James Madison University.

McClave, J.T., Dietrich, Franck H. II, Sincich, T. (1997). *Statistics* (7th ed.). Upper Saddle River, NJ: Prentice-Hall, Inc.

Neuohr, J.L. *The effect of solutions of various pH levels on the growth of kudzu (Pueraria thunbergiana).* Paper prepared for chemistry class at Lee Davis High School, Mechanicsville, VA.

Raybourne, K.D. *Heart girth as an indicator of live weight in white-tailed deer (Odocoileus virginianus).* Paper presented at the Virginia Junior Academy of Science, Second Place, Zoology, Harrisonburg, VA: James Madison University.

Shavelson, R.J. (1996). *Statistical reasoning for the behavioral sciences* (3rd ed.). Boston: Allyn & Bacon.

Yates, D.S., Moore, O.S., McCabe, G.P. (1999). *The Practice of Statistics: TI-83 Graphing Calculator Enhanced.* New York: W.H. Freeman and Company.

Related Web Sites

http://www.dade.k12.fl.us/us1/science/prod03.htm

http://www.mcrel.org/resources/links/index.asp

http://www.eduzone.com/Tips/science/SHOWTIP2.HTM (report section)

http://webster.commnet.edu/mla.htm

http://webster.commnet.edu/apa/apa_index.htm

http://155.43.225.30/workbook.htm

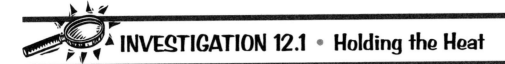

INVESTIGATION 12.1 • Holding the Heat

Question

Which substance makes the best insulator?

Materials for Each Group of 4 Students

3 Aluminum soft drink cans with pull-off tabs
3 Larger cans such as used for vegetables or fruit
3 Thermometers (°C)
900 ml of hot water (80°C)
Insulated container to hold 1000 ml water
Device to heat water such as hot plate or Bunsen burner
Timer
Various insulators such as fiberglass, foam plastic packing foam, sawdust, cotton, air, shredded paper, other commercial products

Safety

● Wear goggles.
● Handle hot materials carefully.

Procedure

Using the general procedures described below, design a class experiment to test the effectiveness of 3 different insulators over a 30-minute period, collecting measurements at regular intervals. Draw the experimental design diagram for the experiment. Write detailed procedures that each lab group can follow. Design a class data table for compiling group data.

● Heat 1 liter of water to a temperature below the boiling point (approximately 80°C).
● Place 3 cm of insulator in the bottom of the larger vegetable can.
● Set the aluminum soft drink can on top of the insulation, being careful to center it.
● Pack the space between the larger vegetable can and the aluminum soft drink can with the insulator.
● Pour a specific amount of water (about 300 ml) into the aluminum soft drink can.
● Insert the thermometer into the water and begin recording the water temperature at regular intervals during the designated time period.
● Compile the data into a class data table.

(continued on the following page)

INVESTIGATION 12.1 ● Holding the Heat *(continued)*

USING TECHNOLOGY ·

1. Load an all-purpose program on your calculator, such as CHEMBIO for Texas Instrument calculators or CASIOLAB for Casio calculators.
2. Use the link cord to firmly attach the calculator to the Data Collector (CBL or EA-100). Turn on both devices.
3. Insert three temperature probes into Channels 1, 2, and 3. Place one probe in each of three aluminum cans.
4. Execute the CHEMBIO or CASIOLAB program on your calculator and follow the screen prompts to enter such information as number of probes, type of probe (thermometers in this case), number of samples and time between samples. If you wish the experiment to run 30 minutes, you should select values that create a total running time of 30 minutes, such as 30 samples 60 seconds apart, or 120 samples 15 seconds apart.
5. When the Data Collector says 'DONE,' follow the directions on the calculator screen to transfer and graph the collected data.

Analyzing Your Data and Reporting Your Findings

1. Summarize your results with an appropriate data table, graph, and paragraph. (Hint: See Table 12.2.)
2. Use a graphing calculator or a computerized statistical program to analyze your data using the correlated *t* test or one-way analysis of variance with repeated measures. For options, see the chapter references.
3. Write a report about the experiment (see Chapters 6, 9, 13).

© 2000 by Kendall/Hunt Publishing Company, Cothron, Giese, & Rezba, *Students and Research.*

INVESTIGATION 12.2 • Practice Makes Perfect

Question

What effect does practice have on ability to perform a task?

Materials for Each Student

Copy of circled numbers handout
Clock

Safety

Be aware of students who have difficulty with hand-eye coordination and assign them an alternative role, if necessary.

Procedure

Read the general procedure outlined below. Plan on a minimum of 30 trials. Draw an experimental design diagram for the experiment and design a class data table for compiling data.

● Keep the handout face down. The handout contains circled numbers from 1 to 59.
● At your teacher's signal, turn the handout over.
● Place your finger on #1, then on #2, and so on until time (30 sec) is called.
● Keep your finger on the last number you reached.
● Record your number in the table below. Turn your handout over.
● Repeat these steps 2 more times for a total of 3 practices.

Number reached in 30 sec

Practice 1	Practice 2	Practice 3
_____	_____	_____

Analyzing Your Data and Reporting Your Findings

1. Summarize your results with an appropriate data table, graph, and paragraph. (Hint: See Table 12.4.)
2. Use a graphing calculator or computerized statistical program to analyze your data using the correlated *t* test or one-way analysis of variance with repeated measures. For options, see the chapter references.
3. Write a report about the experiment (see Chapters 6, 9, 13).

USING TECHNOLOGY ·

1. In the **STAT** mode of your calculator, enter the numbers reached by each person for the first practice in List 1 and for the last practice in List 2. (See Appendix A, *Using Technology*, for additional help in using the graphing calculator).
2. Then select **TESTS** and choose the analysis of variance option (ANOVA) for repeated measures.
3. Enter List 1 and List 2 as the locations of the data, and calculate.

(continued on the following page)

INVESTIGATION 12.2 *(continued)*

INVESTIGATION 12.3 • Fruit Appeal

Question
What factors affect the oxidation of fruit? How do they interact?

General Materials

- Various fruits such as apples, pears, peaches and nectarines
- Paring knives or scalpels for cutting and peeling fruit
- Ways to heat and cool fruit
- Flat dishes to hold slices of fruit (not metal)
- Bowls or other containers to hold liquids (not metal)
- Cutting boards
- Citrus juices such as lemon and orange
- Commercial products for preserving fruit such as ascorbic acid and "Fruit Fresh"
- Various types of plastic wrap, paper, and zip bags
- Other materials to be determined by the lab groups

General Procedures

As long as an apple has its skin, it retains its color and mineral and vitamin content. Once the skin is removed, however, the fruit comes in contact with oxygen and a chemical reaction, oxidation, takes place. Burning occurs with rapid oxidation while rusting and discoloration occur with slow oxidation.

To prevent oxidation you need to keep oxygen away from the substance; for example, you can smother a fire, paint metal, or use an antioxidant. One type of antioxidant, ascorbic acid, combines with the oxygen and creates a protective colorless covering of oxidized material. Thus, the fruit retains its color, rather than turning brown. Because oxidation is a chemical reaction, it is affected by such factors as temperature, concentration, and surface area.

Use classroom, library, or Internet resources to learn more about the factors that affect discoloration of fruit. Then, identify *two or three variables that you think will interact* to affect the oxidation of fruit. As small groups, or as a class, design an experiment to test your hypotheses.

Draw an experimental design diagram for the experiment, develop a set of procedures, and design a data table for compiling group/class data. (Hint: See Tables 12.5, 12.6, 12.8.)

Conduct the experiment and compile the data.

Analyzing Your Data & Reporting Your Findings

1. Summarize your results with appropriate data tables, graphs, and paragraphs. (Hint: See Tables 12.7 and 12.10.)
2. Use a graphing calculator or computerized statistical program to analyze your data using two- or three-way analysis of variance. For options, see chapter references.
3. Write a report about the experiment (see Chapters 6, 8, 9, 13).

USING TECHNOLOGY ·

1. Use the **STAT** mode of your calculator to enter the resulting data from your experiment. (See Appendix A, *Using Technology*, for additional help in using the graphing calculator).
2. Determine the appropriate type of graph for your data and select a scatter plot, histogram, or boxplot to display your data. Graph the data.
3. Using other features of your calculator, perform 1- or 2-variable statistics, predict new values from equations and lines of best fit, or conduct an inferential test as appropriate for the data you collected from your experiment.

© 2000 by Kendall/Hunt Publishing Company, Cothron, Giese, & Rezba, *Students and Research*.

INVESTIGATION 12.4 ● Collapsing Bridges

Question

How is bridge thickness related to collapsing point?

General Materials

Copier paper, letter size   Pennies (all post 1983)
Scissors   Supports such as books or blocks of wood

Procedure

1. Each lab group should cut 3 sheets of letter size copier paper in half length-wise to create 6 long strips of paper.
2. Fold up each end of the paper 3 cm (this adds some rigidity to the paper).
3. Make a bridge by supporting 1 strip of paper between 2 books or other objects of similar height. Position the books 15 cm apart so that the paper bridge will span this distance. Center the paper strip over the span with the folded ends pointing upward.

4. Place a penny in the center of the bridge. Continue to stack pennies until the bridge collapses. Record the number of pennies in the table below.
5. Place a second strip of paper under the original strip to form a 2 layer bridge. Repeat Steps 2–4.
6. Repeat Steps 2–5 using a total of 3, 4, 5, and 6 layers of paper.
7. Enter all lab group data into a class data table. Find the mean (average) number of pennies required to collapse the layered bridges.

Lab Group Data: Number of Pennies Required to Collapse Bridges of Various Thickness					
Layers: 1	2	3	4	5	6
Pennies: _____	_____	_____	_____	_____	_____

Analyzing Your Data and Reporting Your Findings

1. Using data, make a line graph of number of layers versus the number of pennies. Draw a line-of-best-fit. Use the line to predict the number of pennies required to collapse a bridge made of 7 and 10 layers. Make bridges and test your predictions.

(continued on the following page)

INVESTIGATION 12.4 • Collapsing Bridges *(continued)*

Using the class data, make a line graph. Draw a line-of-best fit. Use the line to predict the number of pennies required to collapse bridges made of 7 and 10 layers. Make bridges and test your predictions.

In which line graph do you have the greatest confidence? Why?

2. For the class data, determine the slope of the line and the y intercept. Write a mathematical equation to communicate the relationship between the independent variable (number of bridge layers) and the dependent variable (number of pennies). Remember that the general form of a linear equation is y = ax + b.

Use the equation to predict the number of pennies required to break bridges of other thickness, such as 0.5 layers, 2.5 layers, 7.5 layers, and 12.0 layers. How do the values obtained from the equation and from the graph compare?

3. Write a paragraph summarizing your findings about the relationship between the number of bridge layers and collapsing point

USING TECHNOLOGY ·

1. In the **STAT** mode of your calculator, enter the number of bridge layers in List 1 and the number of pennies in List 2. (See Appendix A, *Using Technology*, for additional help in using the graphing calculator.)
2. In setting up your graph, select scatter plot as your graph type and List 1 for your x-values and List 2 for your y-values. Graph the data.
3. Examine the trend of the data and calculate a line of best fit by performing the appropriate regression (equation) analysis. For example, if the general trend of the data appears to be straight, you may wish to calculate a linear regression. Copy the calculated values to an empty Y=, and graph the equation.
4. To predict the number of pennies it would take to collapse a bridge made of layers that were not measured, press **Trace**, then the 'up' arrow and finally the left or right arrow keys to trace along the best fit line to see the predicted y value (number of pennies) for any desired x-value (bridge layers). Depending on the brand of your calculator you may have to adjust the 'Window' to predict or extrapolate beyond the minimum or maximum x-values.

Extending Your Learning

What other variables do you think affect the strength of bridges? Design experiments to determine the relationship between these variables and breaking point.

Practice

Part I

For each scenario, draw an appropriate experimental design diagram. Some experiments may be represented by more than one design.

1. Larry read that acid rain caused painted surfaces to fade faster. From the literature review, he also found that amount of sunlight and average temperature change also affected paint fading. He established a series of experiments to determine the effect of the three variables on paint fading. Five samples (3 cm × 3 cm) of wood, painted with Brand X Yellow, were exposed to each of the following conditions:

    A. pH of acid rain (pH = 2, 4, 6);
    B. ultraviolet light of 5, 10, 15 units;
    C. day-night temperature change of 10°, 20°, 30°C

2. Lou read that bees were attracted to certain colors. She wondered whether crickets also had a color preference. She divided an aquarium into three sections containing three different colored dishes (red, blue, yellow) with mustard seeds (2 g) in each dish. She placed 30 crickets into the aquarium. She observed the number of crickets in each section at the end of 30, 60, 90, and 120 minutes. She recorded the number of grams of mustard seeds consumed at the end of 120 minutes. She was careful to place the aquarium so that the amount of sunlight was the same in all areas.

3. Nitrifying bacteria on the roots of kudzu are partially responsible for the vine's phenomenal growth rate. Because the bacteria are pH sensitive, Julie Neurohr investigated the influence of altered pH on the growth rate of stems, roots, and leaves. Three hundred kudzu cuttings were allowed to establish themselves in soil for 30 days. The width of roots, stems, and leaves were measured prior to planting. The cuttings were subdivided into groups of 50 cuttings; each subgroup was washed with a solution having a pH of 4, 5, 6, 7, 8, or 9. Each subgroup received 1 L of solution every three days and were grown under comparable light and temperature. For six weeks, stem and leaf width were recorded. At the conclusion of the experiment, plant cuttings were removed from the soil and the root width was measured.

4. Vanessa investigated the effect of two different reward schedules on the learning rates of cats. For three weeks, she allowed five cats to walk a maze and recorded the completion time (maximum of 10 minutes). During the next 3 weeks, she rewarded the cats on a regular schedule when they successfully completed the maze. Each cat's completion time was recorded. A three week rest was allowed, after which she rewarded the cats on a random schedule when they walked the maze. Again, times for completion were recorded; Brand X cat treats were used as a reward. The maze, brand of treat, and interval since the last feeding remained the same.

5. Mary's family encountered conflicting evidence on the effectiveness of various brands of water softening units. Before purchase, Mary investigated the effectiveness of Brands X, Y, and Z with water prepared to three levels of hardness (50 mg/L, 100 mg/L, and 150 mg/L). The hardness, hardness reduction, and increase in sodium concentration after treatment were determined. Ten samples of each type of water were exposed to each softener.

(continued on the following page)

Practice *(continued)*

6. Wildlife biologists typically determine weight of deer with scales. In field conditions, a simpler technique is desirable. Kelly Raybourne investigated the feasibility of chest girth measurements as a substitute. From authorized big game checking stations east and west of the Blue Ridge Mountains, data were secured on sex, age, weight, and chest girth. Across all subpopulations, the relationship between chest girth and weight was investigated. The relationship between chest girth and weight for each subgroup of deer (male and female, young and old, eastern and western Blue Ridge) was also analyzed.

7. Courtney Lynn knew that fish were adversely affected by thermal, phosphate, and acidic pollution. In general, each of these factors negatively influenced respiration, as indicated by a heightened respiratory rate (gill beats/minute). In the experiment, Courtney measured the change in gill beats/minute after 12 hours of exposure to various conditions. In the experiment, she used 120 fish divided into 24 groups, each containing 5 fish. She used three different independent variables. For the first pH, she used three levels (6.5, 5.5, 4.5). For temperature, she used 10°C, 20°C, 30°C, and 40°C. Phosphate concentrations were 0 and 3 ppm. Draw an experimental design diagram and describe the main effects and interactions she could explore.

Part II

Using techniques described in this chapter, determine whether the research hypotheses given in the following scenario are supported by the data.

TABLE 12.12   Data for Weed Killers and Grass Concentration

Concentration	Time				
	8:00 A.M.	**10:00 A.M.**	**12:00 P.M.**	**2:00 P.M.**	**4:00 P.M.**
40%	20	19	10	10	15
	19	18	8	9	17
	20	16	9	11	16
	20	17	11	9	14
	20	16	10	10	15
20%	19	13	4	5	10
	18	12	5	6	9
	16	14	6	7	11
	17	11	4	6	8
	15	12	4	5	7
10%	18	8	3	7	9
	15	7	2	6	7
	16	6	3	7	8
	15	9	4	5	9
	14	8	2	6	6

(continued on the following page)

Practice *(continued)*

1. John read that the biorhythms of organisms influence metabolism of drugs; thus, side effects of drugs can be minimized by prescribing lower concentrations at peak times. Generally, Brand X weed killer is more effective in higher concentrations and when applied early in the day. Because the weed killer acts by interfering with water uptake during photosynthesis, John hypothesized that concentration and time of application would interact to determine maximum effectiveness. John subdivided a field into 15 plots, each containing 5 grass plants. He applied 10 percent, 20 percent and 40 percent X (recommended) at 5 times (8:00 A.M., 10:00 A.M., 12:00 P.M., 2:00 P.M., and 4:00 P.M.). After two weeks, the number of dead plants per plot were recorded (see Table 12.12).

© 2000 by Kendall/Hunt Publishing Company, Cothron, Giese, & Rezba, *Students and Research.*

CHAPTER 13

Preparing Formal Papers

© 1998 PhotoDisc, Inc.

Objectives

- State reasons why a review of the literature should be conducted before you begin an independent research project.
- Describe important information to include in a review of the literature.
- Write a formal review of the literature for an independent research project.
- Identify the elements of a scientific research paper—title page, abstract, introduction, methods and materials, results, discussion-conclusion, and bibliography.
- Write a scientific research paper for an independent science project.
- Use a structured format to teach students to write a scientific research paper.
- Assist individual students in writing appropriate paragraphs of results and conclusions for investigations involving complex experimental designs.

National Standards Connections

- Formulate and revise scientific explanations and models using loci and evidence (NSES).
- Recognize and analyze alternative explanations and models (NSES).
- Design a statistical experiment to study a problem, conduct the experiment, and interpret and communicate the outcomes (NCTM).

Scientists write formal papers to establish the rationale for a proposed research study, as well as other papers to report the findings of the research study. The rationale for the study is established in a review of the literature that includes important background information on the variables, an analysis of prior research on the topic, and a statement of the research problem. Students' research designs and understanding of underlying scientific principles will be improved by requiring a written review of the literature prior

to finalizing the experimental design, procedures, and collection of data. Scientists typically report research through concise research papers. The format of these papers, which is exemplified by articles in scientific journals, includes a title, abstract, introduction, methods-materials, results, discussion-conclusion, and bibliography. Most science competitions require that students report their research in this format. In reporting their research, students will use information from the literature review to write the introduction, to discuss the results, and to formulate a conclusion. In this chapter, procedures for writing a review of the literature and a scientific research paper are summarized and an example of a student research paper is provided.

REVIEW OF THE LITERATURE

Reviews of the literature typically consist of three major components—general background information, an analysis of prior research, and a problem statement. Other components include the title page and a bibliography. Because scientists do not consistently adhere to a single style manual, no particular manual is recommended. However, to maximize interdisciplinary learning, science teachers should use the format required by their school's English department. If science competitive events in your area require a particular style manual, you will need to use that manual for both the literature review and the science research paper. For the review of the literature, other issues such as length of the paper and typed versus handwritten papers should be determined by the teacher.

General Background Information

In this component, students report information on the independent and dependent variables, the specific subject of the study, and any specialized procedures. Because documented informational papers are not taught in most English curricula until grades 10–12, writing a review of the literature overwhelms many students. Techniques for assisting students with initial library research, including the documentation of sources and the preparation of note cards, are described in Chapter 7, *Using Library Resources*. Typical school and library collections, combined with those of community agencies, provide sufficient resources. Thus, it is reasonable to require all students to write a literature review containing general background information on their variables. Structured guidelines that help students determine the essential information to collect and that provide an outline for reporting the literature review are necessary. (See Table 13.1 *How to Write a Review of the Literature* on page 202.)

Begin by having students broadly categorize the variables in their research projects as **plant, animal, protist, matter, energy, process/procedures, or behavior.** Help students learn traditional scientific disciplines that correspond to each category—plant (botany), animal (zoology), matter (chemistry), energy (physics), and earth's processes (geology, meteorology, and so on). Knowledge of discipline names will assist students in locating pertinent information. These seven categories include the majority of variables and subjects investigated by beginning researchers. Appropriately written titles of research projects provide numerous examples and practice opportunities for students to classify variables into the categories.

Title: **The Effect of Aspirin on the Rooting of Kudzu Vines**

IV: Aspirin (Matter—Chemistry)

DV: Rooting of Kudzu Vines (Plant—Botany)

Title: **The Effect of Rotation on the Crystallization of Zinc, Copper, and Lead Sulfate**

IV: Rotation (Energy—Physics)

DV: Crystallization of Zinc, Copper, and Lead Sulfate (Matter—Chemistry)

Title: **The Effect of Three Methods of Terracing on Erosion of Field Soils**

IV: Methods of Terracing (Process/Procedure—Environmental Science and Agricultural Science)

DV: Erosion of Field Soils (Matter—Earth Science and Agricultural Science)

Title: **The Effect of Magnetic Fields on Budding of Yeast**

IV: Magnetic Field (Energy—Physics)

DV: Budding of Yeast (Protist—Botany or Microbiology)

Title: **The Effect of Weekly Stress on the Aggressive Behavior of Toll Attendants**

IV: Weekly Stress (Behavior—Psychology)

DV: Aggressive Behavior of Toll Attendants (Behavior—Psychology).

The broad categories into which a student's research variables fall determine the critical information that the student should find and report. Brief descriptions of important information to be located have proven helpful in focusing students' library research and in providing a structured outline for writing this component of the literature review. For example, a student could use the following descriptions to locate important information about aspirin; a form of matter; and the rooting of kudzu vine, a plant.

Independent Variable, Aspirin, Form of Matter: Briefly describe the substance's chemical and/or common name, formula, physical properties, chemical properties, method of production, and uses. Provide detailed information on its characteristics—chemical and physical properties—that are directly applicable to the study. If you have more than one form of matter, such as aspirin and vinegar, provide this information for each; then, describe similarities and differences of the substances. Based upon the information, predict how the forms of matter will act in the experiment.

Dependent Variable, Rooting of Kudzu Vine, Plant: Provide the plant's common name, scientific name, and classification. Briefly describe the plant's (a) habitat; (b) anatomy including root, stem, leaf, and flower; (c) physiology including how it obtains needed materials, moves, eliminates waste, responds to stimuli, and so on; (d) life cycle, and (e) behaviors or responses to the environment. Provide detailed information in areas that are directly applicable to the study, such as structure and growth of roots. If you have more than one type of plant, such as kudzu and sweet potato vines, provide information on each; then, describe similarities and differences of the plants. Based on the information, predict how the plant(s) will act in the experiment.

Using information obtained through the above process, a student can write the general background portion of the literature review. Brief descriptions of critical information to include

about forms of energy, animals, protists, processes, and behavior are included in Table 13.1. By combining the appropriate categories of questions, adequate structure is provided for most beginning researchers. Teacher assistance can be directed to students with unusual research topics.

Analysis of Prior Research

In this component of the literature review, students describe related research studies. They also ana lyze the studies and establish the implications for their proposed research projects. Although this component is the heart of a professional scientist's literature review, students may be limited by the extent of their scientific knowledge and by the resources of local libraries. Teachers may omit this section for young or inexperienced researchers; however, it should be required of experienced older students, especially if college libraries are available to them (see Table 13.1).

Techniques for assisting students to take notes from scientific journals and abstracts are described in Chapter 7. Critical information to be recorded about each research study includes purpose/hypothesis, experimental design, brief description of procedures, major findings, conclusion, and recommendations for further research. Students can write a succinct description of pertinent research studies from such notes.

Because analyzing and evaluating critical information from several research studies require higher-order thinking, more direct teacher assistance is needed for students to successfully compare research studies. Begin by having students list areas of agreement and disagreement among the studies. Ask students to propose an explanation for any differences among the findings of the studies. Prompt explanations by questioning students about variations in the materials or procedures used in the studies. For example, you might ask: "Were the same types of plants used in the studies?" "Do you think monocots and dicots would respond the same?" "What did you learn about aspirin production in your background reading that might explain differing results with various brands of aspirin?" When students have identified areas of differences, ask what research studies could be conducted to resolve conflicting findings or to address unanswered questions. Ask

TABLE 13.1   How to Write a Review of the Literature

Before you begin the literature review, identify the independent and dependent variables for your research study and follow your teacher's instructions for classifying the variables into the following broad categories.

Animal	Behavior	Matter	Process/Procedure
Energy	Plant	Protists	

Title: Write a sentence that identifies the independent and dependent variables for the investigation.

General Background Information: Describe important characteristics of the independent and dependent variables. Use the following descriptions to help focus your library research and to provide an outline for summarizing the background information.

Animal: Provide the animal's common name, scientific name, and classification. Briefly describe the animal's (a) habitat; (b) anatomy including information on the skeletal, muscular, circulatory, nervous, digestive, and excretory systems; (c) physiology including how it obtains needed materials, moves, eliminates wastes, responds to stimuli, and so on; (d) life cycle; and (e) behaviors or responses to the environment. Provide detailed information in areas that are directly applicable to the study. If you have more than one animal, provide this information on each animal; then, describe similarities and differences of the animals. Based on the information, predict how you think the animals will act in your experiment.

Behavior: Briefly describe the type of behavior, the primary factors influencing the behavior, the value of the behavior, and methods for describing the behavior. Identify critical factors to consider in selecting the sample for the study and for designing the procedure. If you have more than one type of behavior, provide this information on each type of behavior; then, describe similarities and differences. Based on the information, predict how you think the subjects will behave in your experiment.

Energy: Briefly describe the form of energy including how it is produced, measured, transformed into other kinds of energy, and interacts with matter. Cite common examples of the form of energy found in nature or produced by humans. If you have more than one form of energy, provide this information on each form; then, describe similarities and differences of the energy forms. Based on the information, predict how you think the forms of energy will act in your experiment.

Matter: Briefly describe the substance's chemical and/or common name, formula, physical properties, chemical properties, method of production, and uses. Provide detailed information on characteristics that are directly applicable to the study. If you have more than one form of matter, provide this information on each substance; then, describe similarities and differences of the substances. Based on the information, predict how you think the forms of matter will act in your experiment.

Plant: Provide the plant's common name, scientific name, and classification. Briefly describe the plant's (a) habitat; (b) anatomy including root, stem, leaf, and flower; (c) physiology including how it obtains needed materials, moves, eliminates waste, responds to stimuli, and so on; (d) life cycle; and (e) behavior or response to the environment. Provide detailed information in areas that are directly applicable to your study. If you have more than one type of plant, provide this information on each; then, describe similarities and differences of the plants. Based on the information, predict how you think the plants will act in your experiment.

Process/Procedure: Describe the purpose of the process, the major steps, where it occurs, and how it relates to the experiment. If more than one process is involved, provide this information on each process; then, describe similarities and differences of the processes. You may also include pros and cons of each process. Based on the information, predict how you think the processes will affect your experiment.

Protists: Provide the protist's common name, scientific name, and classification. Briefly describe the protist's (a) habitat; (b) anatomy; (c) physiology including how it obtains needed materials, moves, eliminates waste, responds to stimuli, and so on; (d) life cycle; and (e) behavior or response to the envi-

(continued on the following page)

TABLE 13.1 (continued)

ronment. Provide detailed information in areas that are directly applicable to the study. If you have more than one protist, provide this information on each; then, describe similarities and differences of the protists. Based on the information, predict how you think the protists will act in your experiment.

Analysis of Prior Research (Optional): Briefly summarize scientific research studies directly related to your study. Include the purpose, procedures, major findings, and recommendations for further study. If you review more than one study, describe similarities and differences among the research studies. Suggest research studies that need to be conducted to resolve differences or to address unanswered questions.

Statement of the Problem: Describe the rationale, purpose, and hypotheses for the investigation. Use three questions to guide your writing of the introduction:

- Why will you conduct the experiment? (Rationale)
- What do you hope to learn? (Purpose)
- What do you think will happen? (Hypothesis)

Bibliography: List all books, papers, journal articles, and communications cited in the paper. Follow the prescribed bibliographic style manual precisely.

Special Instructions: Follow the teacher's instructions for the length of the paper, format of the paper, style manual, and deadline for submission.

© 2000 by Kendall/Hunt Publishing Company, Cothron, Giese, & Rezba, *Students and Research.*

the students how their proposed study will address these areas, thus expanding an understanding of the topic. When students have successfully identified similarities, differences, and potential research questions, have them write a paragraph that summarizes their analysis.

Statement of the Problem

The literature review ends with a succinct summary of the purpose and rationale for the study and the research hypotheses (see Table 13.1). In the simple scientific reports described in Chapter 6, the problem statement served as the introduction for the report. As with simple reports, students can use three questions to guide their writing of the problem statement.

- Why will you conduct the experiment? (Rationale)
- What do you hope to learn? (Purpose)
- What do you think will happen? (Hypothesis)

FORMAL RESEARCH PAPERS

Rules vary with the competitive event. If students plan to enter research papers in a competitive event, they must know and comply with the rules of the competition. Because papers that do not comply will almost always be disqualified, it is critical that student papers follow the specified guidelines. Most competitive guidelines specify an absolute maximum length. Regulations stating the size of the paper, spacing, margins, and the number of characters per inch are also very precise and will vary. Because looks count, print the final version of the paper on a letter-quality printer. Printing by dot matrix printers should be avoided and, in fact, is sometimes prohibited. Subject to specific guidelines, most formal scientific research papers include a title, abstract, introduction, methods and materials, results, and a discussion-conclusion. Other components include the title page, bibliography, acknowledgements, and appendix. General guidelines for writing these components are summarized in Table 13.2 *How to Write a Scientific Research Paper.* Modify these guidelines as needed to comply with your school's or the competition's requirements.

Initially, students will benefit from reading student research papers and from discussing specific examples of the components outlined in Table 13.2. For this purpose, a high school junior's project on the effect of gibberillic acid on the closing speed of the Venus flytrap is provided. During the discussion, emphasize that students know how to write three of the major report components—the methods and materials, results, and discussion-conclusion—and have essential information for the introduction in their review of the literature. Minor components, such as the title page, abstract, bibliography, acknowledgements, and appendices are easily added (see Chapters 4, 6, and 9).

TABLE 13.2   How to Write a Scientific Research Paper

Title Page

Include the name of the project, your name, the teacher's name, the class, and the date. Follow guidelines provided by your teacher for the format of the page and for additional information required for the competition, for example, school's and parents' names and addresses, hours spent on project, and category.

Abstract

Write a concise summary of your project that includes the problem, hypothesis, procedures, principal results, and conclusions. Do not exceed 250 words.

Introduction

From the review of the literature, summarize information essential for understanding the research project. Include only critical background information on the independent and dependent variables and research studies that directly relate to the research problem. Establish a strong rationale for the study by emphasizing unresolved issues or questions. Conclude by stating the purpose of the study and the research hypotheses.

Methods and Materials

In paragraph form, describe the materials and procedures used to conduct the study. Step listings are not acceptable. Provide sufficient detail to allow a reader to repeat the study. Include precise descriptions of the sample, any apparatus that was constructed or modified for the study, and methods of data collection.

Results

Present the data collected in the experiment in tables and graphs; summarize the data in narrative form. Include statistical analysis of the data. Do not include raw data; if necessary, the raw data can be placed in the appendix. Include only information collected during the study.

Discussion-Conclusion

Restate the purpose of the study, the major findings, and support of the hypothesis by the data. Focus on interpretation of the results. Compare findings with other research; propose explanations for discrepancies. Be sure to provide appropriate literature citations. In addition, make suggestions for procedural improvements and recommendations for further study.

Bibliography

Lists all books, papers, journal articles, and communications cited in the paper. Follow the prescribed bibliographic style manual precisely.

Acknowledgements

Credit assistance received from mentors, parents, teachers, and other sources. As directed by your teacher, include statements by mentors that certify the precise nature of your work. Forms may also be required for human or other vertebrate experimentation.

Appendix

Include critical information that is too lengthy for the main section of the paper, such as raw data, additional tables and graphs, copies of surveys or tests, and diagrams of specialized equipment.

Special Instructions

Precisely follow your teacher's or the competition's guidelines for maximum length of the paper, typing format, use of color or photographs, deadline for submissions, and style format. With competitions, failure to comply may mean automatic rejection.

© 2000 by Kendall/Hunt Publishing Company, Cothron, Giese, & Rezba, *Students and Research*.

STUDENT RESEARCH PAPER

The Effect of Gibberellic Acid on the Closing Speed of the Venus Flytrap
by
Kimberly P. Bryant
Patrick Henry High School, Ashland, VA

Abstract

The Venus flytrap is a small carnivorous plant possessing bilobed leaves in which it captures its prey. The closing of these leaves around a suitably-sized nitrogen-containing object takes place in two distinct phases, one in which the lobes of the leaf snap shut, trapping the prey, and one in which the lobes slowly squeeze together to facilitate digestion. It is believed that the Phase I reaction is brought about by changes in turgor pressure after a trigger hair in the leaf is stimulated, but that the Phase II reaction is an actual growth response. This experiment was designed to test the hypothesis that treating the Venus flytrap with gibberellic acid, a growth hormone, would increase the speed of the Phase II reaction. Ten Venus flytraps were watered with 1 ppm gibberellic acid solution, while ten plants were watered with distilled water every other day for ten days; at the end of this time, the speed of both closing reactions was tested using five leaves from each plant. Though the mean time of both phases was shorter in the experimental group, the results of the t tests showed that only the data from Phase II were significant at the 0.05 level of significance. The research hypothesis that the Phase II closing speed of the Venus flytrap would be increased by the application of gibberellic acid was therefore supported.

Introduction

The Venus flytrap, *Dionaea muscipula,* a member of the *Droseraceae* family, is a carnivorous plant found in damp areas in the eastern part of North Carolina. The roots are small and are used mainly for the absorption of water by the plant; the Venus flytrap obtains most of its nutrients from the insects it traps in its bilobed leaves. Three small triangularly placed filaments are on the inner surface of each lobe of the leaf, perpendicular to the main surface but capable of bending over. When these filaments are stimulated, the lobes of the leaf close together, leaving a small pocket-like enclosure between them. Here the plant secretes digestive enzymes that consume insects or other nitrogen-containing matter. Leaves that have been stimulated to close but that contain no nitrogenous matter generally open within 24 hours, but leaves that close over nitrogenous matter may not open for several days. After two or three digestions the leaf usually dies or is too sluggish to trap any more prey, but leaves may open and close many times if they fail to trap any digestible matter (Darwin, 1972).

Carnivorous plants were for the most part dismissed as exaggerated traveller's tales until Charles Darwin published *Insectivorous Plants* in 1874; this book sparked interest in the phenomenon and gave rise to many experiments on Venus flytraps. Darwin himself performed several experiments by which he determined that the plant is a discerning gourmand, rejecting bits of wood or wax or thick-shelled beetles in favor of ants and spiders. He also determined that the spikes at the margins of the leaves allow very small insects to escape but trap larger insects, saving the plant's powers of digestion for worthwhile prey (Darwin, 1972). One Dr. Curtis in the late nineteenth century in North Carolina discovered that the plants could be successfully maintained on a diet of beef, but that high-protein cheese seemed to cause indigestion and resulted in the death of the leaf (Emboden, 1974). J. M. MacFarlane found in 1872 that two stimuli are required for the closing of the trap; either the same filament must be touched twice or two filaments must be touched (Lloyd, 1976). Darwin had previously believed that only one stimulus was necessary but had noticed that a very light stimulus might not provoke the closing of the leaf (Darwin, 1972).

F. J. F. Meyen made the first attempt to explain the mechanism by which the Venus flytrap closes in 1839, but his explanation that the leaves closed by way of a spring-like mechanism was inadequate. Later researchers hypothesized that the turgor of cells in the leaf must be a factor in closing. Darwin noted that closure appeared to take place in two parts; in the first, the leaf responds quickly to close itself approximately, and in the second, the leaf's lobes press slowly together and the outer surface of the lobes expands (Lloyd, 1976). J. Burdon-Sanderson believed that changes in turgor were the only factors in this reaction, but in 1877, A. Batalin's experiments suggested that actual growth takes place in the leaf during the process of closing and digestion (Lloyd, 1976). It is now thought that the second phase of closing, called the narrowing phase, is caused by a growth spurt; the outer surface grows and forces the lobes inwards, and when the leaf reopens the inner surface grows, forcing the lobes outwards (Slack, 1979).

The gibberellins were discovered in 1926 by E. Kurosawa, a Japanese plant pathologist studying *bakanae,* a disease in which rice seedlings become tall and spindly. It was soon established that the disease was caused by fungus *Gibberella fujikuroi,* and in 1935 T. Yabuta named the active factor in cultures of *G. fujikuroi* "gibberellin." This research was unavailable to the Western world until about 1950, when American and British research teams began trying to purify gibberellin from filtrates of the fungus. In 1954–55 a new compound was isolated from the filtrate, and it is now known as gibberellic acid (GA), $C_6H_{22}O_6$. Since then it has been established that gibberellins occur naturally in a great many species of plants, and at least 52 different gibberellins have been isolated, all having the same basic chemical skeleton (Moore, 1979).

GA works with other plant growth hormones such as the auxins to control cell elongation. The application of GA can even induce normal growth in genetically dwarfed plants. It is also an active factor in such plant growth systems as fruit-set, bolting, and flowering (Devlin, 1975). Although a great deal of experimentation has been conducted using GA, the way in which this growth hormone works upon the plant is not yet fully understood.

Because it is thought that the second phase of trap closing in Venus flytraps is a growth response and gibberellic acid has been shown to induce growth, it is possible that the application of GA may affect the closing speed of the Venus flytrap in some way, perhaps increasing the speed of the second phase of the closing reaction. The purpose of this experiment, then, is to ascertain whether GA affects the closing speed of Venus flytraps. It is unlikely that the results of this experiment, whatever they be, will have any lasting benefits for humanity besides, perhaps, the satisfaction of curiosity; but, after all, one of the primary purposes of science itself is the satisfaction of human curiosity.

Materials and Procedures

The materials used for this experiment were 20 Venus flytraps, a plant light, 1.5 liters of 1 ppm gibberellic acid solution, 1.5 liters of distilled water, a pair of rubber gloves, a 100-milliliter graduated cylinder, tweezers, a stopwatch, and shredded roasted turkey.

The Venus flytraps were set up under the plant light in two groups: the control and experimental groups. On every other day for 10 days, half of the plants were watered with 15 milliliters of distilled water and half of the plants were watered with 15 milliliters of gibberellic acid solution. All testing was done on the eleventh day. The position and reactions of 5 leaves on each plant were recorded; a small piece of shredded turkey was placed in each leaf before the trigger hairs were stimulated, and the Phase I closing time and speed were recorded for each leaf. While Phase I experimentation continued, frequent observations for signs of Phase II closing were made. The time of complete Phase II closing for each leaf was also recorded. The mean time for each closing phase was calculated for both the experimental and control groups, and *t* tests were performed on each set of data at the 0.05 level of significance.

Results

Data Table, Graph, and Statistical Table A show the effect of 1 ppm solution of gibberellic acid on the Phase I closing speed of Venus flytraps. The mean Phase I closing speed of the plants to which gibberellic acid was added (1.7 sec) was less than that of the control group (1.8 sec). There was a greater variation among the control group (range = 2.7 sec) than there was among the experimental group (range = 1.9 sec). Substantial variation existed in both the control group (standard deviation = 0.48) and in the experimental group (standard deviation = 0.46). A t test was used to test the null hypothesis of no significant difference at the 0.05 level of significance. The calculated t value ($t = .83$) was not significant at the 0.05 level of significance, and the null hypothesis, stating that no significant difference would exist between the groups, was therefore accepted.

DATA TABLE A
Effect of GA on Phase I Closing Speed (sec)

Type of Descriptive Data	Control 0 ppm GA	Experimental 1 ppm GA
Mean	1.8 sec	1.7 sec
Minimum	1.1	0.9
Maximum	3.8	2.8
Range	2.7	1.9
Number	50	50

GRAPH A
Effect of GA on Phase I Closing Speed (sec)

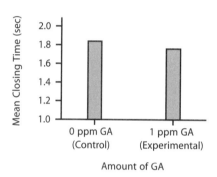

STATISTICAL TABLE A
Effect of GA on Phase I Closing Speed (sec)

Mean difference	0.10 sec
Control standard deviation	0.48
Experimental standard deviation	0.46
Significance level	0.05
Standard *t*-value	2.01
Obtained *t*-value	0.83
Not Significant	

Data Table, Graph, and Statistical Table B show the effect of 1 ppm solution of gibberellic acid on the Phase II closing speed of Venus flytraps. The mean Phase II closing speed of the plants to which gibberellic acid was added (90.2 min) was less than that of the control group (101.2 min). There was a greater variation among the control group (range = 218.0 min) than there was among the experimental group (range = 167.0 min). Substantial variation existed in both the control group (standard deviation = 55.5) and in the experimental group (standard deviation = 46.5). A t test was used to test the null hypothesis of no significant difference at the 0.05 level of significance. The calculated t value (t = 7.68) was significant at the 0.05 level of significance, and the null hypothesis, stating that no significant difference would exist between the groups, was therefore rejected.

DATA TABLE B
Effect of GA on Phase II Closing Speed

Type of Descriptive Data	Control 0 ppm GA	Experimental 1 ppm GA
Mean	101.2 min	90.2 min
Minimum	27.0	21.0
Maximum	245.0	188.0
Range	218.0	167.0
Number	50	50

GRAPH B
Effect of GA on Phase II Closing Speed

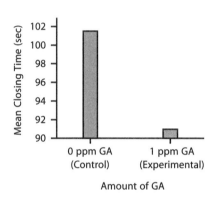

STATISTICAL TABLE B
Effect of GA on Phase II Closing Speed

Mean difference	11.0 min
Control standard deviation	55.5
Experimental standard deviation	46.5
Significance level	0.05
Standard t-value	2.01
Obtained t-value	7.68
Significant	

Conclusion

This experiment was designed to test the effects of a 1 ppm solution of gibberellic acid on the speed of both closing phases of the Venus flytrap's leaves. It was hypothesized that because the second closing phase is thought to be a growth response, the application of gibberellic acid would increase the speed of the reaction; however, the first phase of closing, which is basically a change in turgor pressure, would not be affected. It was found that the mean time for each phase was less in the group of plants treated with gibberellic acid. These data appeared not to support the hypothesis concerning Phase I but to support the hypothesis concerning Phase II; that is, the data suggested that the speed of both reactions was increased by the application of gibberellic acid.

However, results of *t* tests completed on the data showed that there was a significant difference between the control and experimental groups only with the second phase of closure; the research hypothesis was supported by the data. The first phase of closure was unaffected by gibberellic acid, whereas the speed of the second phase of closure was increased. The data may have been affected somewhat by uncontrolled environmental variables such as temperature and humidity; this experiment could have been improved by keeping the Venus flytraps in a climatorium, where more outside variables could be controlled. Using more plants would also increase the accuracy of the data. For further research, extensive records could be kept on each leaf so that individual variations in closing speed could be correlated and accounted for; then the effect of gibberellic acid on the Venus flytrap could be established more firmly.

ORIGINAL REFERENCES CITED

Acid explanation of Venus flytrap spring. *Science News,* January 15, 1983, p. 41.

Acid flux triggers the Venus flytrap. *New Scientist,* March 3, 1983, p. 582.

The Allure of Carnivorous Plants. *New Scientist,* February 6, 1986, p. 32.

Baker, N. R., Davies, W. J., & Ong, C. K., editors. *Control of Leaf Growth.* Cambridge: Cambridge University Press, 1985.

Boxer, Sarah. The subtlest assassins. *The Sciences,* May/June 1983, pp. 7–8.

Cell growth causes plant to shut its trap. *New Scientist,* January 8, 1981, p. 72.

Darwin, Charles. *Insectivorous plants.* New York: AMS Press Inc., 1972.

Davies, P. J., editor. *Plant hormones and their role in plant growth and development.* Dordrecht, The Netherlands: Martinus Nijhoff Publishers, 1987.

Devlin, Robert M. *Plant physiology.* New York: Litton Educational Publishing, 1975.

Emboden, William A. *Bizarre plants: Magical, monstrous, mythical.* New York: Macmillan Publishing Company, Inc., 1974.

Lloyd, Francis Ernest. *The carnivorous plants.* New York: Dover Publications, Inc., 1976.

Moore, Thomas C. *Biochemistry and physiology of plant hormones.* New York: Springer-Verlag New York Inc., 1979.

Scott, Tom K. *Hormonal regulation of development II: The functions of hormones from the level of the cell to the whole plant.* Wurzburg, Germany: Springer-Verlag Berlin Heidelburg, 1984.

Slack, Adrian. *Carnivorous plants.* Cambridge, Massachusetts: The MIT Press, 1979.

Wallace, Robert A., King, Jack L., & Sanders, Gerald P. *Biology: The science of life.* Glenview, IL: Scott, Foresman, and Company, 1986.

Williams, Stephen E., Bennett, Alan B. Leaf closure in the Venus flytrap: An acid growth response. *Science,* December 10, 1982, pp. 1120–22.

ACKNOWLEDGEMENTS

I would like to thank Mrs. Eleanor Tenney, who gave me the idea for this project, Mrs. Pamela Gentry, who helped me put it together, and Dr. Julia Cothron, who explained the statistics to me. I would also like to thank my mother, who drove for miles to find the best deal in Venus flytraps and then patiently answered the questions of all the people who wanted to know why anyone would want so many strange-looking plants, and my father, whose first-rate skills at on-the-spot school-project carpentry provided a home for the Venus flytraps.

USING TECHNOLOGY ·
SURFING THE WEB FOR CARNIVOROUS PLANTS

More than a decade ago, Kimberly Bryant conducted her study of the effect of Gibberellic Acid on the closing speed of the Venus flytrap. Then, middle and high school students had limited access to scientific references, unless a college/university library was nearby. Obtaining up-to-date references was much more difficult than today, because the Internet has revolutionized students' abilities to obtain information.

If your students were interested in carnivorous plants, such as the Venus flytrap, they would need to use much more recent references than the ones cited in Kimberly Bryant's paper. Except for classical studies, such as the ones by Darwin and Kurosawa, scientists generally do not cite references that are more than ten years old. Fortunately, your students have search engines and the electronic world at their fingertips. Here are some handy websites we found on the topic:

- Botany Encyclopedia of Plants (www.botany.com)
- Botany Libraries—Harvard University Herbaria (www.herbaria.harvard.edu/Libraries/libraries)
- Botany Smithsonian Databases (http://nmnhwww.si.edu/botany/database.htm)
- Botanical Society of America (www.botany.org)
- Carnivorous Plants: Cultivation (www.geocities.com/RainForest/Vines)
- The Carnivorous Plant (www.sarracenia.com)
- International Carnivorous Plant Society, Inc. (www.carnivorousplants.org)
- Internet Directory for Botany (www.botany.net)

Check out the sites we have identified. Then, use your ingenuity to find others. Did you learn more about the effect of gibberellic acid on Venus flytraps? Did you find other interesting questions to investigate?

PRACTICING AND REINFORCING SKILLS

Suggestions for integrating the writing of simple reports into the science curriculum are described in Chapter 6, *Writing Simple Reports*. The strategies of practicing one component at a time and using cooperative learning strategies can also be applied to writing a review of the literature and a formal research report.

Through laboratory activities, students can practice the three major components of a literature review. For example, if students were investigating the influence of light on the photosynthetic rate of algae, they could use high school biology, chemistry, or physics texts to secure basic information about light, photosynthesis, and algae. Then, they could write a brief literature review consisting of general background information on variables and a problem statement. The typical chemistry topic, models of atomic structure, could provide a valuable opportunity to practice the analysis of prior research component of a literature review. Students could summarize the major theories of atomic structure and compare the similarities and differences. Similarly, a physics experiment on the characteristics of light could be preceded by a review of the wave and particle theories of light. Suggestions for assessing students' reviews of the literature are provided in *Evaluating for Success 9: Writing a Review of the Literature.*

Students can also cooperatively write scientific research papers on classroom experiments. One student can write the introduction, a second the methods and materials, a third the results, and a fourth the discussion-conclusion. Results may include descriptive or inferential statistics as appropriate. Suggestions for evaluating scientific research papers are provided in *Evaluating for Success 10: Writing a Scientific Research Paper* in Chapter 17.

REFERENCES

Clyne, D. (1998). *Plants of prey*. Wisconsis, WI: Gareth Stevens.

D'Amato, P. (1998). *The savage garden: Cultivating carnivorous plants*. Berkeley, CA: Ten Speed Press.

Gibaldi, J. (1998). *MLA style manual and guide to scholarly publishing* (2nd ed.). New York: Modern Language Association of American.

Bryant, K.P. *The effect of gibberellic acid on the closing speed of the Venus fly-trap*. Paper presented at the meeting of the Virginia Junior Academy of Science, First Place in Botany Division and Botany Award, Richmond, VA: Virginia Commonwealth University.

Chicago Editorial Staff. (1993). *The Chicago manual of style: The essential guide for writers, editors, and publishers* (14th ed.). Chicago: University of Chicago Press.

Gentle, V. (1996). *Bladderworts: Greasy cups of death (Bloodthirsty Plants)*. Milwaukee, WI: Gareth Stevens.

Gentle, V. (1996). *Pitcher plants: slippery pits of no escape (Bloodthirsty Plants)*. Milwaukee, WI: Gareth Stevens.

Gentle, V. (1996). *Venus Fly Traps and waterwheels: Spring traps of the plant world (Bloodthirsty Plants)*. Milwaukee, WI: Gareth Stevens.

Juniper, B.E., Robins, R.J., & Joel, D.M. (1989). *The carnivorous plants*. Orlando, FL: Academic Press.

Publication manual of the American Psychological Association (4th ed.). (1994). Hyattsville, MD: American Psychological Association.

Simons, P. (1996. August). When a carnivore is not a carnivore. *New Scientist*, 151, p. 16.

Turabian, K. L. (1996). *A manual for writers of term papers, theses, and dissertations* (6th ed.). Chicago: University of Chicago Press.

Related Web Sites

http://www.dade.k12.fl.us/us1/science/prod03.htm

http://www.mcrel.org/resources/links/index.asp

http://www.eduzone.com/Tips/science/SHOWTIP2.HTM (report section)

PART THREE

Management Strategies for Classroom and Independent Research

✓ Successfully integrate inquiry into your curriculum and promote strong school-community ties:

- overcome curriculum constraints
- assess student achievement
- effectively schedule research
- communicate with parents

Encouraging Parental Support

Objectives

- Describe the essential information to communicate about research to parents.
- Describe ways to involve parents and the community in student research projects.
- Develop letters for communicating information about research to parents.
- Plan activities to actively involve parents and the community in student research.

Substantial research evidence exists to show that students at all grade levels perform better when parents support teachers and become involved in their children's learning. Most parents were their children's first science teacher—answering questions, exploring the outdoors, and allowing their children to work with tools and to watch things mix and change in the kitchen. **Science teachers can once again encourage parental involvement in science through middle and high school by keeping parents informed and by asking for their assistance.** Letters to parents about various aspects of independent science research are included in this chapter. Science teachers can use, as models, those letters that are appropriate to their situation.

INITIAL INVOLVEMENT OF PARENTS

The first two letters suggest ways to initially involve parents in experimental science (see Letters 14.1 and 14.2). Parent Letter 14.1 describes a unique form of homework in which students are asked to involve their family in a simple science experiment that illustrates many of the experimental design concepts being learned in class. A set of directions for an experiment using paper helicopters is included. The behavior of pendulums, the reaction of backyard isopods (pillbugs) to light, and the color changes of a natural pH indicator like red cabbage broth to various items in the kitchen represent other suitable take-home activities in physics, biology, and chemistry. Any simple and safe activity that involves manipulating a variable and observing the response can be used. The second letter invites parents to a special back-to-school program sponsored by the science department. During such a program a rationale for independent research, the appropriate role of parents in student projects and a schedule of events can be discussed.

PARENT LETTER 14.1 • Family Involvement

Dear Parents:

In science we are learning how experiments are designed and conducted. I thought you might enjoy an example of the kind of activities and experiments we are doing in class. These class activities are designed to develop in students the skills and concepts they will need to independently conduct their own experiments later in the year. Attached to this letter are the directions for conducting simple experiments that I hope you and other members of your family will enjoy doing with your son or daughter.

As you participate in the activity, it would be good practice for your son or daughter to explain to you the following basic concepts of experimental design and to identify them in the experiment:

● independent and dependent variables
● hypothesis
● constants
● control
● repeated trials

Explanations of these concepts are included below.
Thank you for your cooperation.

Sincerely,

Basic Concepts of Experimental Design

Hypothesis: A prediction about the relationship between the variables that can be tested, for example, **If** you increased the amount of light plants receive, **then** the growth of the plants will also increase.

Variables: The factors that could vary or be changed in an experiment, for example, amount of light, fertilizer, water, type of plant, growth of plants, quality of fruit produced.

Independent Variable: The variable that is purposefully changed by the experimenter, such as amount of light.

Dependent Variable: The variable that responds, such as amount of growth of the plants.

Constants: All factors that were kept the same and have a fixed value, for example, plants received the same amount of water and fertilizer, the same type of plant was used, and weather conditions were the same for all plants.

Repeated Trials: The number of experimental repetitions, objects, or organisms tested, for example, for each amount of light 10 plants could be used. Repeated trials are used to reduce the effects of chance or random errors; this increases confidence in the results.

Control: The group that is used as a standard for comparison in an experiment; it is often the group that received no treatment, but it can also be a group that is designated by the investigator as a comparison group, for example, the group of 10 plants that received no light or the group of 10 plants that received the "normal" amount of light could be designated as the control.

© 2000 by Kendall/Hunt Publishing Company, Cothron, Giese, & Rezba, *Students and Research*.

FAMILY ACTIVITY 14.1 • Experimenting with Helicopters

Cutting the Strip

The attached sheet can be cut into three strips that can then be folded to form individual helicopters. Cut the sheet so you have three strips that look like *Diagram 1*.

Cutting and Folding Each Strip

Cut along the dotted lines and then fold along the solid lines beginning with the line between sections A and B.

1. Fold the section labeled A behind the section B.
2. Fold section C behind section B.
3. Fold section D behind section B as well.

Complete the helicopter by folding blade X in one direction and blade Y in the opposite direction *(see Diagram 2)*.

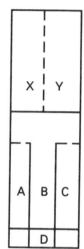

Diagram 1

Activity Directions

Hold your helicopter up high and drop it. Do this several times and observe the results.

● Do all helicopters behave the same?
● Do they all spin in the same direction?
● What can you do to the helicopter to change the spin direction?
● What happens if you refold the blades in the opposite direction?

In this experiment, what was the independent variable? What was the dependent variable?

Diagram 2

Try Another Experiment

What do you think will happen to the helicopter if you increase its weight by adding a paper clip? Try it several times, and describe how the helicopter behaved.

Repeat the experiment, increasing the number of paper clips (for example, 2, 3, 4, 5 paper clips). What happens as you increase the weight?

● What is your independent variable?
● What is your dependent variable?
● Describe your control for this experiment.
● What factors were held constant in this experiment?

© 2000 by Kendall/Hunt Publishing Company, Cothron, Giese, & Rezba, *Students and Research*.

Helicopter Master

X	Y	X	Y	X	Y

A	B	C	A	B	C	A	B	C
	D			D			D	

RESEARCH AT HOME

Students often select topics that involve the use of living organisms, chemicals, or hazardous procedures. If student research is to be conducted at home, parental permission must be obtained. Letters should provide parents with information regarding their supervisory role as well as the humane care of living organisms and the potential hazards of each chemical and procedure to be used.

When the cost of organisms, chemicals, and other materials are the responsibility of the parents, specific information on these charges should be included. Each of the three permissions letters (14.3, 14.4, and 14.5) should also contain an appropriate return portion for the parents to grant or deny permission.

INSUFFICIENT PROGRESS

Not every student views timelines and due dates as seriously as science teachers would like. When a reminder is in order, a letter addressed to both the student and the parents can be sent home and returned with a parent signature. An alternate ap-

---

PARENT LETTER 14.2 • General Overview and Meeting Invitation

Dear Parents:

As your son's or daughter's science teacher, I want to share some of the goals we have as a science department for students this year. One of these goals is to involve students in independent scientific research. Our science department believes that all students should have the opportunity to investigate a problem of their choice. With assistance from several sources, each student will select a topic and conduct an experiment to study some aspect of that topic. Most students will need to do some of their work at home. Some will actually conduct their experiments at home and all will need to work in libraries to research their topic. A schedule of events and a timeline for completing various parts of the experimental project are attached. Please review them with your son or daughter.

Students who choose to enter their experiment in one or more competitions will need your support and encouragement in preparing for and in attending those competitive events. These events include science fairs at various levels and the Junior Academy of Science. More information will be provided later.

You are invited to a special evening meeting of the science department for an overview of our program. During the meeting we will model techniques students may follow in selecting and conducting a science project. The appropriate roles of parents in their son's or daughter's project will be discussed. As a special treat, student projects from last year will be presented and displayed. Following this presentation, refreshments will be served. We hope you will attend with your son or daughter. The meeting will begin at 7:30 P.M. on Thursday, September 21st, and will end by 9:00 P.M.

The science department has always enjoyed the support and participation of our students' parents. I look forward to working with you to give your son or daughter the best science education possible. If you have any questions, please call me at ____ *phone* ____ .

Sincerely,

© 2000 by Kendall/Hunt Publishing Company, Cothron, Giese, & Rezba, *Students and Research.*

Photograph by Arthur Schwieder; print by Eleanor and Wilton Tenney.

PARENT LETTER 14.3 • Live Organisms at Home Permission

Dear Parents:

As you may know, your son or daughter has chosen an experimental project that involves living organisms and he or she has asked the science department to order the required organisms. There will be a charge of $_____*amount*_____ for them. Please send a check payable to the school by *date*. Research grants to cover the cost of materials are available; students may obtain an application form from the science department. Please contact me at _____*phone*_____ if you have any questions.

Because of our concern that all living materials be humanely treated, we are asking that you let us know whether you are willing to have these organisms in your home and that you will supervise the care of these organisms while they are in your home. Instructions for the proper care of the organisms will be provided.

Please complete the bottom part of this letter and have your son or daughter return it by _____*date*_____.

Sincerely,

_____ I give permission for these organisms to be in our home and I will supervise their care.

_____ I do not want these organisms in our home.

Parent's Signature _____ Student's Signature _____

© 2000 by Kendall/Hunt Publishing Company, Cothron, Giese, & Rezba, *Students and Research*.

PARENT LETTER 14.4 • Chemicals at Home Permission

Dear Parents:

Your son or daughter has chosen an experimental project that requires the use of one or more chemicals. Listed below are those chemicals, their potential hazards, and the precautions that should be taken.

Chemical Hazards/Precautions/Disposal

_____ *(Enter specific information)* _____

We are concerned that all chemicals be used safely and request your help in supervising their use at home. Please contact me at _____ *phone* _____ if you have questions.

Please complete the bottom part of this letter and have your son or daughter return it by _____ *date* _____.

Sincerely,

_____ I have read the potential hazards and precautions above and will supervise the proper use of these chemicals in our home.

_____ I do not want these chemicals in our home.

Parent's Signature _____ Student's Signature _____

© 2000 by Kendall/Hunt Publishing Company, Cothron, Giese, & Rezba, *Students and Research*.

PARENT LETTER 14.5 • Hazardous Procedures at Home Permission

Dear Parents:

Your son or daughter has chosen an experimental project that includes conducting one or more procedures that are potentially hazardous. Listed below are these procedures, their potential hazards, and the precautions that should be taken.

Procedures   Hazards/Precautions

_____ *(Enter specific information)* _____

We are concerned that all potentially hazardous procedures be conducted carefully and request your help in supervising them at home. Please contact me at _____ *phone* _____ if you have questions.

Please complete the bottom part of this letter and have your son or daughter return it by _____ *date* _____.

Sincerely,

_____ I have read the potential hazards and precautions above and will supervise the proper use of these procedures in our home.

_____ I do not want these procedures conducted in our home.

Parent's Signature _____ Student's Signature _____

© 2000 by Kendall/Hunt Publishing Company, Cothron, Giese, & Rezba, *Students and Research*.

proach is to mail the parents a copy of a reminder letter given to students in class. Whenever students are making insufficient progress on their projects, a similar procedure should be followed to remind students and to inform parents (Letters 14.6 and 14.7).

PARENTAL ASSISTANCE

Although parents can provide assistance in many ways, most teachers particularly appreciate their help in taking students to area libraries. Library research is rarely easy, but it can be less frustrating if relevant library materials can be located. Because few schools have science research collections, the use of nearby college libraries or other scientific library collections may be necessary. After using the strategies suggested in Chapter 7 to provide students with the library skills to successfully conduct a review of related literature, inform parents of the importance of the library research through Parent Letter 14.8.

Parental assistance in reviewing the first draft of students' papers can also be very helpful, especially when a science teacher has a large number of students involved in projects. In Parent Letter 14.9, parents are asked to review their son's or daughter's paper and, in addition, to certify that it is the result of the student's own work.

PARTICIPATING IN COMPETITIONS

Four letters related to student participation in competitive events are included. The first two of these letters (14.10 and 14.11) inform parents that their son's or daughter's project has been recommended for entry in a district science fair or for presentation at the state Junior Academy of Science. A return portion indicating parental permission to submit is necessary, because students will be expected to make a presentation of their project if selected; transportation and out-of-town living expenses may be necessary.

STUDENT/PARENT LETTER 14.6 • Insufficient Progress on Project Topic

Dear Student/Parent:

   This letter is to remind you that an important day is approaching very rapidly. As you know, you have had several weeks in which to decide upon a topic for an independent science research project.

   The due date for your project topic is close at hand; according to my records you are still without an acceptable research project. Remember this is a required project that will increasingly affect your grade in succeeding marking periods. Please make an appointment with me before or after school to discuss project ideas for your very own investigation.

Sincerely,

© 2000 by Kendall/Hunt Publishing Company, Cothron, Giese, & Rezba, *Students and Research.*

STUDENT/PARENT LETTER 14.7 • Insufficient Progress in Conducting Research

Dear Student/Parent:

At the beginning of the year you were given a schedule of events and a timetable for completing various aspects of your experimental project. An additional copy is attached for your review.

Indicated on that timeline are those assignments for which you have made insufficient progress. Failure to make satisfactory progress jeopardizes your completion of this important assignment and your grade in science.

Please make an appointment with me before or after school to discuss your project. I will be happy to help you get back on schedule.

Sincerely,

© 2000 by Kendall/Hunt Publishing Company, Cothron, Giese, & Rezba, *Students and Research.*

PARENT LETTER 14.8 • Library Research

Dear Parents:

Your son or daughter has chosen a science topic to research that should begin with a review of related literature. The quality of students' research can be greatly improved if they have access to library resources beyond those typically available in a school library. Although it is not mandatory to use other libraries, your son or daughter would benefit from the experience of using other library collections.

Your assistance would be greatly appreciated in helping your son or daughter make use of one or more of the library resources available in this area. These include

(List here nearby colleges, universities, and large public libraries, as well as document collections at extension offices, farm bureaus, and local, state, and federal government agencies.)

_____

If you have any questions, please call me at _____*phone*_____.

Sincerely,

PARENT LETTER 14.9 • Parent Review of Draft Research Paper

Dear Parents:

On _____ date _____ the first draft of a paper describing your son's or daughter's science project will be due. Please review his or her paper and certify that to the best of your knowledge the project and the paper are the student's work. The paper should follow the printed organizational instructions presented and discussed in class and should contain the following sections:

- Title Page and Abstract
- Introduction
- Methods and Materials
- Results
- Discussion-Conclusion
- Bibliography
- Acknowledgements
- Appendix if applicable

Both student and parent should sign the Title Page to certify that the work presented is that of the student.

Thank you for your continued support. If you have any questions, please call me at _____ phone _____ .

Sincerely,

© 2000 by Kendall/Hunt Publishing Company, Cothron, Giese, & Rezba, *Students and Research*.

PARENT LETTER 14.10 • Participation in a District/Regional Science Fair

Dear Parents:

Because of the outstanding quality of your son's or daughter's science project, the project has been recommended for entry in the district (regional) science fair.

Students whose projects are submitted must display their projects at the science fair and discuss them with the judges. This year's science fair will be held in _____ location _____ on _____ date _____ . [Note: If participation requires an overnight stay, see Letter 14.11 for additional information to include in this letter.]

Please use the form below to indicate your permission to have your son's or daughter's project submitted for judging in the district (regional) science fair.

Sincerely,

_____ I understand that if my son's or daughter's project is submitted, he or she will be expected to make a presentation of the project in _____ location _____ .

_____ My son or daughter may not go to _____ location _____ ; please do not submit the project for judging.

Parent's Signature _____   Student's Signature _____

When students participate in science competitions, science teachers need as much parental support as possible. Parent Letter 14.12 provides information about an upcoming science competition and also invites parents to attend a session for students to practice their oral presentations. The final letter (14.13) can be used to thank parents for their assistance and to ask them to encourage their son or daughter to write a thank you note to other individuals and organizations who helped with the recent science competition.

Related Web Site

http://www.eduzone.com/
Tips/science/
SHOWTIP2.HTM
(parents section)

PARENT LETTER 14.11 • Participation in a Competitive Event Requiring a Formal Paper

Dear Parents:

Because of the quality of your son's or daughter's science project, the project has been recommended for entry in the _____state_____ Junior Academy of Science (JAS). All students whose projects have been recommended are required to submit a formal paper describing their research by _____date_____.

Students whose papers are selected by the judges must give an oral presentation describing their project at the JAS Meeting. Being selected is an honor and will give your son or daughter the opportunity to interact with other outstanding students from across the state. This year's meeting will be held on the campus of _____college_____ in _____city_____ on _____date_____. Students will spend two nights in a college dormitory and eat their meals in campus cafeterias. The cost of room and board is expected to be about _____amount_____. Financial assistance is available for students in need. Additional information will be provided later.

Please use the form below to indicate your permission to have your son's or daughter's project submitted for judging in the state JAS.

Sincerely,

_____ I understand that if my son's or daughter's paper is selected, he or she will be expected to make a presentation of the project in _____city_____.

_____ My son or daughter may not go to the JAS meeting; please do not submit the paper for judging.

Parent's Signature _____ Student's Signature _____

© 1998 PhotoDisc, Inc.

© 2000 by Kendall/Hunt Publishing Company, Cothron, Giese, & Rezba, *Students and Research*.

PARENT LETTER 14.12 • Information on Upcoming Science Competition and Invitation to Science Night

Dear Parents:

I know how proud you must be that your son's or daughter's science research project has been selected to be presented at the _____*state*_____ Junior Academy of Science (Regional Science Fair). This year's meeting (fair) will be held at the _____*University*_____ in _____*city*_____. Students will leave school by bus on _____*date/time*_____ and return on _____*date/time*_____ . Transportation costs and the registration fee for all students will be paid by the science department. The cost of two nights lodging and six meals will be _*amount*_. We will be staying in college dormitories and having our meals in campus cafeterias. Checks should be made payable to the school and are due by _____*date*_____ . Financial assistance is available for students in need; please call me at _____*phone*_____ for more information.

_____ I will be attending the Science Research Night at school so I can be part of the audience for practicing students.

_____ I would like to serve as a chaperon for the trip. My telephone number is _____.

_____ I would like to attend the presentation part of the meeting in _____*city*_____ ; enclosed is the _*amount*_ registration fee for observers.

_____*Name*_____ has my permission to attend the _____*state*_____ Junior Academy of Science (Regional Science Fair) at _____*location*_____ on _____*date*_____ .

Parent's Signature _____   Student's Signature _____

PARENT LETTER 14.13 • Letter of Appreciation

Dear Parents:

We could not have done it without you! Thank you for your help in making this year's (science competition, such as the _____*state*_____ Junior Academy of Science (Regional Science Fair) such a success. I am very proud of your son's or daughter's accomplishments as I know you are.

On a separate sheet provided to your son or daughter, I have listed the names and addresses of several individuals and organizations who helped us prepare for and attend the recent _____*science competition*_____ . I have asked all my students to express their gratitude by writing a thank you note to at least one person on the list. Would you please encourage them to do so; handing them a stamp might be just the subtle nudge they need. Thanks again for your help.

Sincerely,

····· **C H A P T E R 15** ···········

Assessing with Paper-Pencil Tests

> ### Objectives
>
> ● Successfully answer paper-pencil items on the concepts of experimental design and data analysis.
> ● Use paper-pencil items to assess students' understanding of the concepts of experimental design and data analysis.
> ● Develop paper-pencil items to assess students' understanding of the concepts of experimental design and data analysis.

Most science teachers rate experimentation and laboratory activities as being an important part of science learning. Students, however, tend to rate experimentation and laboratory experiences as much less important than class lectures and text material. Tests contribute to these differing views. To students, the relative importance of specific knowledge is more efficiently communicated by tests than by any other means.

If something is important enough to teach, it is important enough to test. Yet science tests, including lab practicals, seldom contain items that assess student knowledge and skills of experimental design and data analysis. This chapter provides sample items that can be used to directly assess the mastery level of these concepts and skills. You can select and use individual items within your tests. Better still, use these sample items as models for your own development of similar items based on experiments that your students do. Also, any of the problems at the end of the chapters of this book that are not assigned as practice, can be used to assess student learning.

BASIC CONCEPTS
(Chapters 1 and 2)

1. Each of the following terms identifies a component of an experiment. Define each of the following terms: (a) independent variable; (b) dependent variable; (c) constant; (d) repeated trials; (e) control; and (f) hypothesis.

2. Identify the independent and dependent variables in an experiment with the following title: **The Effect of Placing Used Tea Bags Under Rose Plants on the Growth of the Rose Plants.**

_____ Independent variable

_____ Dependent variable

3. Identify the variable that was purposely changed and the responding variable in the hypothesis: **If the kind of dry cell put into a toy car is changed, then the total distance the car will travel changes.**

_____ Purposefully changed variable

_____ Responding variable

4. Match each term in Column II with its definition in Column I.

Column I	Column II
_____ 1. A statement of a possible relationship between the independent and dependent variables.	A. Control
_____ 2. Any factor that is not allowed to change.	B. Independent variable
_____ 3. A group or sample that is used as a standard for comparison.	C. Repeated trials
_____ 4. Used to reduce the effects of chance errors.	D. Constant
_____ 5. The factor in an experiment that is changed on purpose.	E. Dependent variable
_____ 6. The factor in an experiment that responds to the purposefully changed factor.	F. Variable
_____ 7. Any factor in an experiment that changes.	G. Hypothesis

5. In the experimental design diagram shown below, each letter, A through E, represents a component of an experiment. Identify the component that each letter represents.

Title:
Hypothesis:

A			
B	B	B	B
C	C	C	C

D _____
E _____

Letter   Component

A. _____

B. _____

C. _____

D. _____

E. _____

6. In the experimental design diagram shown here, the various components of an experiment on the effectiveness of different gasolines are illustrated. Write a specific example of the designated component in the blanks below.

Sam predicted that Ace gasoline was best.

Brand of gasoline		
Speedy	**Ace**	**Roll-on**
3 round trips (Attica, NY to Downtown Whigham, GA)	3 round trips (Attica, NY to Downtown Whigham, GA)	3 round trips (Attica, NY to Downtown Whigham, GA)

Miles per gallon
Same make, model, and horsepower autos
3 cars traveled together
- same route
- same traffic
- same speed, starts, and stops

Component	Specific Example
Independent variable	_____
Hypothesis	_____
Repeated trails	_____
Constants	_____
Dependent variable	_____
Levels of independent variable	_____

7. For **each** of the following scenarios: (a) construct an experimental design diagram, including a title and a hypothesis; (b) identify the control (if present); and (c) list at least two ways to improve the experiment.

Scenario A, Freezing Water: June wished to determine whether the concentration of salt in water affected how long it takes water to cool. She put 4 identical plastic glasses, each of which contained 225 ml of a different concentration of salt solution into a freezer, for example, 0 percent, 10 percent, 20 percent, 30 percent salt. She recorded the amount of time it took for each solution to cool to a temperature of 3°C.

Scenario B, Bees and Flower Color: Jackie wanted to know whether the color of a flower affected the attraction of bees to the flower. She had many flowers growing in her backyard. She observed the number of the bees that landed on red roses, blue hydrangeas, yellow marigolds, and pink carnations. She observed for 30 minutes. She reported the number of bees that landed on each of the four colors of flowers.

Scenario C, Magnets and Force: Michael bought 4 identical, inexpensive bar magnets and labeled them A, B, C, and D. He put the north end of magnet A into iron filings and measured the mass of the filings the magnet could lift. He dropped Magnet B from a height of 2 meters onto a cement floor. From the same height, he dropped Magnet C twice and Magnet D three times. Using the north end of the magnets, he picked up iron filings. He measured the amount of filings lifted by each magnet on the same scales. Michael was careful to use the same batch of iron filings to test each magnet.

GENERATING IDEAS AND PROCEDURES
(Chapters 3 and 4)

8. Use the **Four Question Strategy** to brainstorm ideas for experiments on *one* of the following topics: **shadows, birdseed, air filters, nonprescription antibiotic ointments.**

   Question 1: What materials are readily available for conducting experiments on
   _____ ?

   Question 2: How do _____ act?

   Question 3: How can I change the set of _____ materials to affect the action?

   Question 4: How can I measure or describe the response of _____ to the change?

9. Using your answers to **Question 8,** draw an **experimental design diagram,** including a title and a hypothesis for *one* experiment that you could conduct.
10. Using your answers to **Questions 8 and 9,** write a **procedure** for the experiment you diagrammed.
11. Write a detailed and precise **procedure** that includes both the sequence of steps to be done and the materials needed for one of the following:

   A. Assume you are standing in front of a sink. State how you could determine the change in the rate at which water comes out of the faucet for each 45 degrees that the faucet handle is turned.
   B. Suppose you wanted to determine the effect of water temperature on the rate at which sugar cubes would dissolve.
   C. A visiting exchange student wants a quarter-pound cheeseburger on her first day in the United States. Explain how she can use a drive-through window at a local fast food restaurant.

CONSTRUCTING TABLES AND GRAPHS
(Chapter 5)

12. For each experiment title listed below, state whether the experiment should be graphed as a **bar** or a **line** graph.

   _____ A. The Effect of the Amount of Fertilizer Used on the Height of Sunflower Plants
   _____ B. The Effect of the Gender of Voters on the Choice of Candidates for President
   _____ C. The Effect of Temperature on the Number of Days It Takes for Mealworms to Complete Their Life Cycle
   _____ D. The Effect of the Degree of Ripeness on the Number of Peaches Sold
   _____ E. The Effect of the Number of Foxes in an Area on the Number of Rabbits in That Same Area

13. Construct a *data table* and an appropriate *graph* for the following set of data:

Paper	Number of New Subscriptions (1989)
The Weekly Press	40,000
The News Herald	42,000
The Florida Gazette	5,000
The Bennington Journal	8,500
The World Yesterday	10,000

14. Construct a *data table* and an appropriate *graph* for the following set of data about Buddy's progress in physical therapy.

Time of Therapy	Number of Steps Buddy Walked
9:10 (start)	0
10:05	3
9:40	3
9:15	4
11:00	5
9:30	5
10:15	5
10:40	5
10:35	6

15. For Graphs A, B, and C below, *describe the relationship* between the variables shown.

Graph A

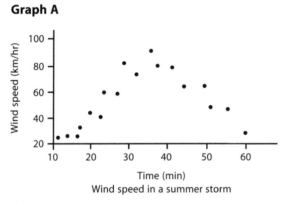

Wind speed in a summer storm

Relationship:

Graph B

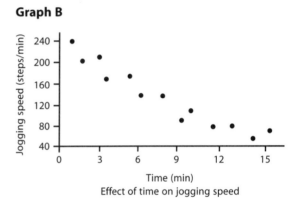

Effect of time on jogging speed

Relationship:

Graph C

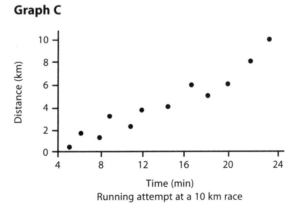

Running attempt at a 10 km race

Relationship:

ANALYZING EXPERIMENTAL DATA
(Chapter 8)

16. For each of the terms listed here, place a **C** in front of each measure of central tendency. Place a **V** in front of each measure of variation within a set of data. Place an **N** in front of each term that is neither a measure of central tendency nor of variation.

_____ range	_____ mean	_____ frequency distribution
_____ sum	_____ number	_____ median
_____ mode	_____ discrete	_____ continuous

17. For each of the following types of data, state the measures of central tendency and variation that should be used.

Type of data	Measures of Central Tendency	Measure of Variation
Nominal	_____	_____
Ordinal	_____	_____
Interval	_____	_____
Ratio	_____	_____

18. Rodney measured the height of 100 three-week-old bean plants in centimeters. For each plant he recorded the color of its leaves. Identify each set of data that Rodney collected. Classify each set as being qualitative or quantitative. Justify your classification of each set of data.

19. Rene studied the number of grams of a deep yellow solid that formed when she varied the amount of 0.1 M lead nitrate solution she combined with an equal amount of 0.1 M potassium chromate solution. Were Rene's data quantitative or qualitative? Justify your answer.

20. Give an example of each type of data listed here.

Type of Data	Example
A. Nominal data	_____
B. Ordinal data	_____
C. Interval data	_____
D. Ratio data	_____

21. For Scenarios A–C answer the questions below.
   A. What type of data is represented by the dependent variable?
   B. What level of measurement is represented by the dependent variable?
   C. What is the most appropriate measure of central tendency?
   D. What is the most appropriate measure of variation?
   E. Construct a table for summarizing the data.
   F. Compute the appropriate measures of central tendency, variation, and number; enter them in the table.

   Scenario A, Inclined Plane: John wanted to determine whether the angle of an inclined plane affected the speed of a toy car. When he set the ramp at 0°, he recorded no motion in each of three trials. With the ramp set at 40°, he recorded trial runs of 2 sec, 3 sec, and 3 sec. With the ramp set at 30°, he recorded three trial runs of 4 sec each. With the ramp set at 20°, he recorded runs of 10 sec, 14 sec, and 12 sec. And with the ramp set at 10°, he recorded runs of 22 sec, 24 sec, and 18 sec.

Scenario B, Flower Color: Julie Ann wondered whether people who bought groceries in the supermarket where she worked purchased one color of roses more than another. At 5 P.M. each Friday in June, she recorded the number of each color of rose sold until she had sold 100 roses. The color of roses sold were red (R), yellow (Y), pink (P), white (W), and lavender (L). On June 1, she sold 24 R, 10 Y, 30 P, 25 W, and 10 L. On June 8, she sold 20 Y, 15 R, 28 P, 25 W, and 12 L. On June 15, she sold 16 W, 29 R, 12 Y, 32 P, and 11 L. On June 22, she sold 13 L, 24 W, 24 P, 18 Y, and 21 R. On June 29, she sold 30 R, 26 P, 10 Y, 25 W, and 9 L.

© 1998 PhotoDisc, Inc.

Scenario C, The Lighted Aquarium: Elizabeth Anne thought that varying amounts of light in different areas of an aquarium would affect the number of brine shrimp that would use an area for their habitat. She arranged the lighting system for an aquarium so that she had 4 areas: brightly lighted = 4, moderately lighted = 3, dimly lighted = 2, and very dimly lighted = 1. She put the lighting over 5 identical aquaria containing brine shrimp. She waited 24 hours. She sealed the areas of each aquarium and counted the number of brine shrimp in each area.

	Amount of lighting			
Aquarium \ Areas	Area 4	Area 3	Area 2	Area 1
1	18	61	17	12
2	136	83	28	13
3	102	75	11	8
4	117	68	21	10
5	110	87	15	12

22. Suppose the following set of data was *qualitative* data; compute the median, the mode, and a frequency distribution.

24	25	20
21	26	19
22	27	18
27	28	27
24	29	23

23. Suppose the set of data in Question 22 was *quantitative* data; compute the mean and the range.
24. All of the following are examples of ratio data *except* the number of

    a. blue corn kernels
    b. wrinkled peas
    c. degrees Celsius
    d. seashell fossils

25. The sum of all the values in a set of interval data divided by the total number of values describes the

    a. mean
    b. median
    c. mode
    d. variance

26. For a set of ordinal data, which measure of central tendency would best describe the central or most typical value?

 a. mean
 b. median
 c. mode
 d. range

27. Given a set of nominal data, which measure of central tendency should be used?

 a. mean
 b. median
 c. mode
 d. range

28. The spread of numerical scores on a particular science test could best be described using the

 a. median
 b. mean
 c. range
 d. frequency distribution

29. The variation in the number of each color T-shirt in a set of T-shirts would best be described by the

 a. median
 b. mean
 c. range
 d. frequency distribution

30. Which of the following would provide the most information in determining the meaning of a student's score of 145 points on a test?

 a. the variation and median
 b. the range and mean
 c. the range and mode
 d. the frequency distribution and the mode

Use Jim's data to answer questions 31–35.

Jim did an experiment to determine how the angle of a ruler-rubber band launcher affected the flight of a rubber band. He placed the bottom edge of a ruler flat on a desk (0°). He placed a rubber band on the top edge of the ruler so that the front end of the rubber band just caught on the end of the ruler. He stretched the rubber band 4.5 cm. He released the rubber band 20 times and each time he measured the distance from the front of the ruler to the place where the rubber band came to rest. He raised the front of the ruler so that the ruler made a 10°, a 30°, and a 50° angle with the table. At each angle he repeated the procedure 20 times. Below are Jim's data.

Angle of Ruler-Rubber band launcher									
0°					10°				
390	364	374	362	361	368	371	354	352	359
360	353	368	359	332	351	381	373	361	374
387	341	357	372	362	384	365	378	385	351
375	382	373	347	354	364	386	382	368	371
30°					50°				
441	424	441	426	448	380	392	426	424	410
443	407	448	449	430	423	424	389	418	422
431	445	449	432	441	416	388	392	401	404
392	436	433	424	419	399	413	386	385	384

31. Make a stem and leaf plot for the data for each angle.
32. For which angle size of the ruler-rubber band launcher does the distance data form a J distribution?
    a. 0°
    b. 10°
    c. 30°
    d. 50°

33. For which angle size of the ruler-rubber band launcher does the distance data form a U-shaped distribution?
    a. 0°
    b. 10°
    c. 30°
    d. 50°

34. For which angle size of the ruler-rubber band launcher does the distance data form a normal distribution?
    a. 0°
    b. 10°
    c. 30°
    d. 50°

35. For which angle size of the ruler-rubber band launcher does the distance data form a rectangular distribution?

    a. 0°
    b. 10°
    c. 30°
    d. 50°

Use the following data to answer questions 36–43.

The Effect of Different Average Size Soil Particles on the Water Retention (ml) of the Soil															
Silt Loam (medium grain size)								Sandy Loam (large grain size)							
35	36	37	36	43	39	39	42	32	34	38	37	32	36	37	30
40	37	39	39	39	42	39	42	30	34	38	40	28	31	30	29
36	36	37	40	41	40	39	41	38	36	36	36	34	36	36	31
40	38	41	37	41	41	37	41	31	35	39	31	37	37	32	35

36. What is the median value of the amount of water retained in the sandy loam? _____ ml
37. What is the Q_3 value of the amount of water retained in the sandy loam? _____ ml
38. What is the Q_1 value of the amount of water retained in the sandy loam? _____ ml
39. For which of the two soils is the data skewed?

    a. sandy loam
    b. silt loam
    c. neither
    d. both

40. What is the standard deviation in the water retention data for the silt loam? _____

    Use a t test to determine the statistical significance of the difference of the water retention data for sandy and silt loams.

41. Calculate t. $t =$ _____
42. Is the difference in water retention of the two types of soils significant? _____
43. If the difference is significant, at what level is it significant?

DETERMINING STATISTICAL SIGNIFICANCE
(Chapter 11)

44. Ms. Hanney a school nurse, randomly selected for a research study 10 students with head lice from her school district, which had over 100 students with head lice. Which group of students would be considered the: (a) sample, (b) sampled population, (c) target population?

45. Use the sets of students described in the following diagram to match the terms in Columns I and II.

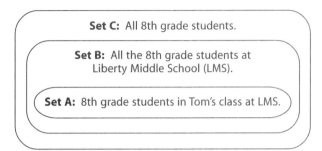

Set C: All 8th grade students.

Set B: All the 8th grade students at Liberty Middle School (LMS).

Set A: 8th grade students in Tom's class at LMS.

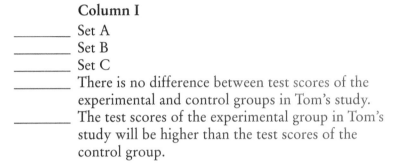

Column I	Column II
_____ Set A	1. Null hypothesis
_____ Set B	2. Target population
_____ Set C	3. Research hypothesis
_____ There is no difference between test scores of the experimental and control groups in Tom's study.	4. Sample
_____ The test scores of the experimental group in Tom's study will be higher than the test scores of the control group.	5. Sampled population

Use the following scenario and data sets to answer Questions 46 and 47.

The Banana Scenario: Anthony worked in the produce department of a grocery store in the afternoons. He wondered how much the price and ripeness of bananas affected the sale of the fruit. He hypothesized that when the price of bananas was low, fruit sales would increase. To test this hypothesis, he collected data for 5 days on the total number of pounds of bananas sold when priced at 20, 30, and 40 cents per pound. Anthony's data on the influence of price on sales are shown in Data Set 1. Anthony also hypothesized that the ripeness of bananas would affect sales, with the greatest number of sales occurring when bananas were yellow. To test this hypothesis, Anthony determined the number of pounds of green, yellow, and brown-black bananas sold on Saturday; these data are shown in Data Set 2.

**Data Set 1
Effect of Price on
Banana Sales (pounds)**

Trial	Cost/Pound ($)		
	.20	**.30**	**.40**
1	20	20	5
2	25	30	10
3	20	40	5
4	15	25	15
5	30	20	10

© 1998 PhotoDisc, Inc.

**Data Set 2
Effect of Ripeness on
Banana Sales (pounds)**

Degree of ripeness		
Green	**Yellow**	**Brown/black**
15	20	7
10	15	5
10	30	8
15	25	8
5	20	5

46. For Data Set 1 of Anthony's data, complete the following questions.
    A. What is the research hypothesis?
    B. What is the appropriate inferential statistical test?
    C. State an appropriate null hypothesis.

D. Calculate the statistical value (show all work).
E. Using the 0.05 level of significance, is the calculated value significant?
F. What are the degrees of freedom?
G. Should the null hypothesis be accepted or rejected?
H. Should the research hypothesis be accepted or rejected?
I. Construct an appropriate data table for displaying both the descriptive and inferential statistics.
J. Write an appropriate paragraph of results.

47. For Data Set 2 of Anthony's data, answer Questions A–J listed in Question 46.

Use the following scenario and data sets to answer Questions 48–50.

The Tylex Scenario: Kim wanted to find out if diluting Liquid Tylex would change the speed at which it would dissolve the material that causes the ring-around-the-tub. Each day for 30 days she submerged the lower half of 12 tiles in used bath water for 30 minutes. She then soaked 3 of the tiles in 250 ml pure Liquid Tylex. She placed 3 more of the tiles in a mixture of 200 ml Tylex and 50 ml water. She placed another 3 of the tiles in a mixture of 150 ml Tylex and 100 ml of water. She placed another 3 of the tiles in a mixture of 100 ml Tylex and 150 ml of water. The final time (sec) it took for the grease ring to disintegrate or disappear and the cloudiness of the resulting solutions (C = Clear; CL = Cloudy; VCL = Very Cloudy; SCL = Slightly Cloudy) were recorded.

Data Set 1: Time for Ring to Disappear (seconds)

250 ml Tylex	200 ml Tylex 50 ml water	150 ml Tylex 100 ml water	100 ml Tylex 150 ml water
165	221	348	511
170	226	413	528
165	229	352	552

Data Set 2: Cloudiness of Tylex/Water Solutions
(C = Clear; CL = Cloudy; VCL = Very Cloudy; SCL = Slightly Cloudy)

Pure Tylex	200 ml Tylex 50 ml water	150 ml Tylex 100 ml water	100 ml Tylex 150 ml water
C	C	CL	VCL
C	SCL	CL	VCL
CL	SCL	CL	CL

48. For the above scenario, answer the following questions:
   A. Construct an experimental design diagram that communicates the independent and dependent variables, constants, control (if present), and repeated trials.
   B. State an appropriate research hypothesis related to the time required to remove the grease ring.
   C. State an appropriate research hypothesis for the cloudiness of the resulting solutions.

49. For Data Set 1 of Kim's data, answer the following questions:
   A. What is the best measure of central tendency?
   B. What is the best measure of variation?

50. For Data Set 2 of Kim's data, answer the following questions:
    A. What is the best measure of central tendency?
    B. What is the best measure of variation?

Use the following scenario and data set to answer Questions 51–55.

Jerry read that exercise makes the mind more alert. He hypothesized that a good mixture of study and exercise would increase test grades on studied materials. He divided 40 high ability math students into two groups of 20 each. Both groups received the same instruction on a kind of math problem that was new to them. Both groups were then given 50 practice exercise problems. Each member of group A did the 5 problems and then 20 situps followed by 5 more of the problems and 20 more situps until the problem set was complete. Group B members did 5 problems, relaxed for 30 seconds (the time Group A took to do 20 situps) and did 5 more problems, relaxed for 30 seconds, and so on, until the 50 problems were completed. After the problems were completed, Jerry tested the students' ability to solve similar problems. His data are summarized here:

	Exercised group (A)	Nonexercised group (B)
Mean ($\bar{x}$) score	55.0	56.0
Standard deviation	7.0	10.0
Number	20	20

Calculated $t = 1.20$ Required $t\,(0.01) = 2.87$

51. Which group of students had the best performance on the test?
    a. Exercised group
    b. Nonexercised group

52. Which group of students exhibited the most variation in learning?
    a. Exercised group
    b. Nonexercised group

53. Ninety-five percent of the students in the exercised group have scores between
    a. 48 and 62
    b. 46 and 66
    c. 41 and 69
    d. 36 and 76

54. The appropriate number of degrees of freedom in determining the required t value is
    a. 38
    b. 40
    c. 20
    d. 222

55. Based on the data, you would
    a. support the research hypothesis
    b. not support the research hypothesis

REFERENCE

Rezba, R.J. et al. (1995). *Learning and assessing science process skills.* Dubuque, IA: Kendall/Hunt Publishing Company.

CHAPTER 16

Assessing with Rating Sheets

© 1998 PhotoDisc, Inc.

Objectives

- Distinguish between formative and summative evaluation.
- Justify the use of formative evaluation to evaluate components of a student research project.
- Describe the essential components of student research to be assessed with rating sheets.
- Use rating sheets to assess students' scientific research skills.

Scientific methodology epitomizes ongoing or formative evaluation. Through a review of the literature, researchers refine the problem statement, the methods and materials, and the data analysis to be used in an investigation. Initial experiments provide feedback that leads to new and improved experimental designs and procedures. On completion of an experiment, a final report is written that ends with yet another example of formative evaluation, recommendations for further study. Through peer review, the scientific community renders final or summative evaluation. In school, the typical student experiences primarily summative evaluation. Quality independent research projects, however, depend on formative evaluation. Techniques for providing this ongoing feedback include teacher, peer, parental, and self-review on both a formal and an informal basis. Checklists or rating sheets can enhance this process by focusing attention on key elements.

Suggested rating sheets included in this chapter focus on both the basic and advanced prin-

ciples of experimental design. For a given rating sheet, several criteria for evaluation are listed along with suggested point values to use in assigning a grade if desired. Each rating sheet is correlated with chapters that describe how to teach the concepts and the importance of the criteria listed in evaluating student work. The rating sheets provide for self, peer/family, or teacher evaluation. They can be used to provide either formative or summative evaluation of student progress. The 11 rating sheets, provided in Tables 16.1 to 16.11, are summarized here.

1. Designing and Generating Experiments
2. Describing Experimental Procedures
3. Constructing Tables and Graphs
4. Writing a Simple Report
5. Using Library Resources
6. Collecting Experimental Data
7. Analyzing and Communicating Data: Descriptive Statistics
8. Analyzing and Communicating Data: Inferential Statistics

9. Writing a Review of the Literature
10. Writing a Scientific Research Paper
11. Presenting Scientific Research

Designing and Generating Experiments: A two-part rating sheet for the assessment of a student's understanding of the basic concepts of experimental design and the use of the *Four Question Strategy* (see Table 16.1).

Describing Experimental Procedures: Criteria for assessing a student's ability to develop procedures, based on the Four Question Strategy and to finalize procedures for implementation (see Table 16.2).

Constructing Tables and Graphs: A three-part rating sheet for the assessment of a student's ability to construct data tables, line graphs, and bar graphs (see Table 16.3).

Writing a Simple Report: Criteria for assessing a student's ability to write a simple report including the title/introduction, experimental design, procedures, results through data tables and graphs, and a conclusion (see Table 16.4).

Using Library Resources: Criteria for assessing a student's ability to properly identify sources, take notes, organize and submit information gathered from general sources, scientific research, technical procedures, and interviews (see Table 16.5).

Collecting Experimental Data: Criteria for assessing a student's ability to collect raw data and to submit it according to a timeline; also provided is a sample format for requesting revisions of project and extensions of deadlines (see Table 16.6).

Analyzing and Communicating Data—Descriptive Statistics: A four-part rating sheet for the assessment of a student's ability to construct data tables and graphs and to write paragraphs of results and a conclusion (see Table 16.7).

Analyzing and Communicating Data—Inferential Statistics: A four-part rating sheet for the assessment of a student's ability to conduct an inferential statistical test, to construct a data table containing inferential statistics, and to write paragraphs of results and a conclusion (see Table 16.8).

Writing a Review of the Literature: Criteria for assessing a student's ability to provide general background information on the independent and dependent variables, to analyze prior research (optional), to state the research problem, to display good writing skills, and to follow the general format for a formal paper (see Table 16.9).

Writing a Scientific Research Paper: Criteria for assessing a student's ability to write an introduction, methods and materials, results, and a discussion-conclusion, to display good writing skills, and to follow the general format for a formal paper (see Table 16.10).

Presenting Scientific Research: Criteria for assessing a student's ability to communicate the research, to answer questions, and to present research effectively; optional criteria are provided for visual displays and oral presentations (see Table 16.11).

TABLE 16.1   Designing and Generating Experiments

Criteria/Value	Self	Peer/Family	Teacher
1 Evaluating for Success			
Name _____ Period _____ Date _____			
Part One—Basic Concepts of Design (100 points)			
Title (5)			
Hypothesis (5)			
Independent variable (10)			
Levels of independent variable (10)			
Control (10)			
Repeated trials (10)			
Dependent variables (10)			
Operational definition of dependent variable (10)			
Constants (15)			
Experimental design diagram (10)			
Creativity/Complexity (5)			
Part Two—Four Question Strategy (100 points)			
Q1: Readily available materials (30) Excellent list Good list Poor list			
Q2: Action of materials (10) Excellent answer (correct) Good answer (partially correct) Poor answer (incorrect)			
Q3: Ways to vary materials (30) Excellent list Good list Poor list			
Q4: Ways to measure actions (20) Excellent list Good list Poor list			
Creativity of prompt (5)			
Creativity of brainstorming (5)			

Chapter Correlations
1—Developing Basic Concepts
3—Generating Experimental Topics
17—Scheduling Student Research

© 2000 by Kendall/Hunt Publishing Company, Cothron, Giese, & Rezba, *Students and Research*.

TABLE 16.2   Describing Experimental Procedures

		2 Evaluating for Success	
Name _____ Period _____ Date _____			
Criteria/Value (100 points)	Self	Peer/ Family	Teacher
All steps included (30)			
All materials/Equipment included (20)			
Written for one level of independent variable (10)			
Repetitions for repeated trials (10)			
Repetitions for levels of independent variable (10)			
Written in approved format—lists or paragraph (10)			
Spelling/Grammar (5)			
Sentence/Paragraph structure (5)			
Special Comments (See Student Paper)			

Circled items require permission to use living organisms or hazardous chemicals/procedures.

Starred items (*) may be expensive or difficult to obtain; consider alternative materials, community sources, or grant funds.

<u>Underlined items</u> involve vertebrate experimentation; you will need to obtain a mentor or consider alternatives.

Chapter Correlations
4—Describing Experimental Procedures
14—Encouraging Parental Support
17—Scheduling Student Research

© 2000 by Kendall/Hunt Publishing Company, Cothron, Giese, & Rezba, *Students and Research*.

TABLE 16.3　Constructing Tables and Graphs

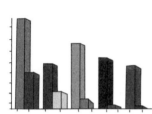

Criteria/Value	Self	Peer/Family	Teacher
3 Evaluating for Success			
Part One—Data Tables (100 points)			
Title (10)			
Vertical column for independent variable (10)			
Title/Unit of independent variable included (5)			
Values of independent variable ordered (10)			
Vertical column for dependent variable (10)			
Title/Unit of dependent variable included (5)			
DV column subdivided for repeated trials (10)			
Dependent variables correctly entered (10)			
Vertical column for derived quantity (10)			
Unit of derived quantity included (10)			
Derived quantity correctly calculated (10)			
Part Two—Line Graphs (100 points)			
Title (10)			
X axis correctly labeled including units (10)			
Y axis correctly labeled including units (10)			
X axis correctly subdivided into scale (15)			
Y axis correctly subdivided into scale (15)			
Data pairs correctly plotted (15)			
Data trend summarized with line-of-best-fit (10)			
Data trend summarized with sentences (15)			
Part Three—Bar Graphs (100 points)			
Title (10)			
X axis correctly labeled including units (10)			
Y axis correctly labeled including units (10)			
X axis correctly subdivided—discrete values (15)			
Y axis correctly subdivided into scale (20)			
Vertical bars for data pairs correctly drawn (15)			
Data trend summarized with sentences (20)			

Name _____ Period _____ Date _____

Chapter Correlations
5—Constructing
　　Tables & Graphs
17—Scheduling
　　Student Research

© 2000 by Kendall/Hunt Publishing Company, Cothron, Giese, & Rezba, *Students and Research*.

TABLE 16.4   Writing a Simple Report

	4 Evaluating for Success		
Criteria/Value	**Self**	**Peer/ Family**	**Teacher**
Title/Introduction (16)			
Correct title (4)			
Rationale (4)			
Purpose (4)			
Hypothesis (4)			
Experimental design (16)			
Name/Levels/Units of independent variable (4)			
Control (4)			
Repeated trials (4)			
Name/Units of dependent variable (4)			
Procedures (12)			
All steps, equipment, and materials included (4)			
Written for one level of independent variable (2)			
Repetitions for repeated trials and levels of IV (2)			
Spelling/Grammar (4)			
Results—data tables (16)			
Labeled vertical column for independent variable (4)			
Labeled vertical column for dependent variable (4)			
Labeled vertical column for derived quantity (4)			
Correct values of IV, DV, derived quantity (4)			
Results—graphs (16)			
Correct label/Unit/Scale for X axis (4)			
Correct label/Unit/Scale for Y axis (4)			
Data pairs correctly plotted (4)			
Data trends summarized (4)			
Conclusion (24)			
Purpose of experiment (2)			
Major findings (4)			
Support of hypothesis by data (4)			
Comparisons/Explanations (4)			
Recommendations—Further Study/Improvement (4)			
Spelling/Grammar (6)			

Name _____ Period _____ Date _____

Chapter Correlations
6—Writing a Simple Report
17—Scheduling Student Research

© 2000 by Kendall/Hunt Publishing Company, Cothron, Giese, & Rezba, *Students and Research*.

TABLE 16.5 Using Library Resources

Criteria/Value	Self	Peer/ Family	Teacher
5 Evaluating for Success			
Name _____ Period _____ Date _____			
Proper Identification of Sources (15)			
Call number/Phone number (2)			
Location/Address (3)			
Correctly documented source (10)			
Note-taking skills (25)			
Accurate information (5)			
Paraphrased words and phrases (10)			
Quotations around authors' words (5)			
Page numbers noted (5)			
Organizational skills (10)			
Required number of sources submitted (5)			
Legible writing (2)			
Cards organized by source and numbered (3)			
Required information (50)—Use appropriate criteria for assignment.			
General source — Topic sentence (10)			
Major points (30)			
Additional references (10)			
Scientific research — Purpose/Hypothesis (5)			
Experimental design (10)			
Procedure (5)			
Major findings/Conclusion (15)			
Areas for further research (10)			
Additional references (5)			
Technical procedures — Name of procedure (5)			
Materials/Equipment/ Availability (15)			
Brief synopsis of steps (15)			
Ability to execute (10)			
Additional references (5)			
Interview — Questions for interview (15)			
Responses to questions (15)			
Reviewing/Editing of notes (10)			
Letters of appreciation (10)			

Chapter Correlations
7—Using Library Resources
17—Scheduling Student Research

© 2000 by Kendall/Hunt Publishing Company, Cothron, Giese, & Rezba, *Students and Research*.

TABLE 16.6   Collecting Experimental Data

		6 Evaluating for Success		
Name _____ Period _____	Date _____			
Criteria/Value (100 points)		**Self**	**Peer/ Family**	**Teacher**
Management of time (25)				
All data submitted on time (25)				
Partial data submitted on time (15)				
No data submitted on time/No approved extension (0)				
Adequacy of progress (25)				
Satisfactory progress (25)				
Partially satisfactory progress (15)				
Unsatisfactory progress (0)				
Collection of raw data (50)				
Quantitative data collected (10)				
Raw data table for quantitative data (10)				
Qualitative data/Observations recorded (10)				
Raw data table for qualitative data (10)				
Sufficient measurements/Observations (5)				
Evidence of experiment being conducted (5)				
Revisions & Extensions				
Request for Revisions or Deadline Extension *(For additional space, use the back of this sheet.)*				
1. What is your project topic?				
2. What revisions do you propose? Why are they necessary?				
3. What part of your data will be turned in on schedule?				
4. When will you turn in the other data?				
Teacher Approval _____ Date _____				

© 2000 by Kendall/Hunt Publishing Company, Cothron, Giese, & Rezba, *Students and Research*.

Chapter Correlations
5—Constructing Tables and Graphs
8—Analyzing Experimental Data
17—Scheduling Student Research

TABLE 16.7   Analyzing and Communicating Data: Descriptive Statistics

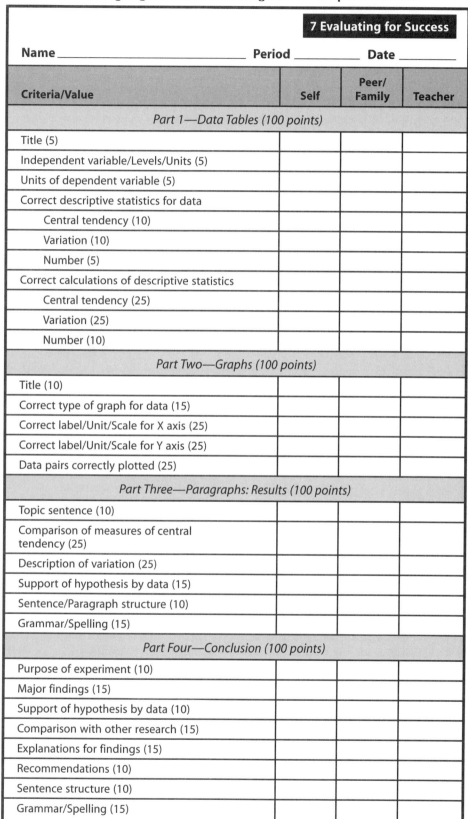

	7 Evaluating for Success		
Name _____ Period _____ Date _____			
Criteria/Value	Self	Peer/ Family	Teacher
Part 1—Data Tables (100 points)			
Title (5)			
Independent variable/Levels/Units (5)			
Units of dependent variable (5)			
Correct descriptive statistics for data			
Central tendency (10)			
Variation (10)			
Number (5)			
Correct calculations of descriptive statistics			
Central tendency (25)			
Variation (25)			
Number (10)			
Part Two—Graphs (100 points)			
Title (10)			
Correct type of graph for data (15)			
Correct label/Unit/Scale for X axis (25)			
Correct label/Unit/Scale for Y axis (25)			
Data pairs correctly plotted (25)			
Part Three—Paragraphs: Results (100 points)			
Topic sentence (10)			
Comparison of measures of central tendency (25)			
Description of variation (25)			
Support of hypothesis by data (15)			
Sentence/Paragraph structure (10)			
Grammar/Spelling (15)			
Part Four—Conclusion (100 points)			
Purpose of experiment (10)			
Major findings (15)			
Support of hypothesis by data (10)			
Comparison with other research (15)			
Explanations for findings (15)			
Recommendations (10)			
Sentence structure (10)			
Grammar/Spelling (15)			

© 2000 by Kendall/Hunt Publishing Company, Cothron, Giese, & Rezba, *Students and Research*.

Chapter Correlations
8—Analyzing
  Experimental Data
9—Communicating
  Descriptive
  Statistics
10—Displaying
  Dispersion/
  Variation in Data
17—Scheduling
  Student Research

TABLE 16.8 Analyzing and Communicating Data: Inferential Statistics

Criteria/Value		Self	Peer/ Family	Teacher
	8 Evaluating for Success			
Name _____ Period _____ Date _____				
Part One—Inferential Statistical Test (100 points)				
Correct null hypothesis (10)				
Correct level of significance (5)				
Correct statistical test (10)				
Correct calculations (30)				
Correct degrees of freedom (10)				
Correct table value for statistic (5)				
Correct interpretation of test—significance (10)				
Correct action about null hypothesis (10)				
Correct action about research hypothesis (10)				
Part Two—Data Tables (100 points)				
Title (10)				
Descriptive statistics (40)				
Name of inferential statistical test (10)				
Comparison of calculated/Table values (20)				
Degrees of freedom (10)				
Significance/Probability level (10)				
Part Three—Paragraphs: Results (100 points)				
Topic sentence (10)				
Comparison of descriptive statistics (25)				
Description of statistical test (15)				
Interpretation of statistical test (15)				
Support for research hypothesis (10)				
Writing/Grammar/Spelling (25)				
Part Four—Conclusion (100 points)				
Purpose of experiment (10)				
Major findings, including statistical test (15)				
Support of research hypothesis by data (10)				
Comparison with other research (15)				
Explanations for findings (15)				
Recommendations (10)				
Writing/Grammar/Spelling (25)				

© 2000 by Kendall/Hunt Publishing Company, Cothron, Giese, & Rezba, *Students and Research*.

Chapter Correlations
11—Determining
    Statistical
    Significance
17—Scheduling
    Student Research

TABLE 16.9   Writing a Review of the Literature

		9 Evaluating for Success	
Name _____ **Period** _____ **Date** _____			
Criteria/Value	**Self**	**Peer/ Family**	**Teacher**
General background information (30–60)			
Independent variable (15–30)			
Dependent variable (15–30)			
Prior research—optional (0–30)			
Description of prior research (10)			
Analysis of prior research (10)			
Questions for future study (10)			
Statement of problem (15)			
Rationale (5)			
Purpose (5)			
Research hypothesis (5)			
Writing (15)			
Logical organization/Effective transitions (5)			
Sentence/Paragraph structure (5)			
Grammar/Spelling (5)			
General format (10)			
Title page (2)			
Bibliography (4)			
Footnotes (2)			
Other requirements (2)			

Comments

Comments: You need to add additional information in the areas that are circled.

Animal/Plant/Protist
Name/Classification
Anatomy
Physiology
Life Cycle
Behavior/Response
Comparisons
Predictions

Behavior
Type
Factors influencing
Value
Methods describing
Sample selection
Comparisons
Predictions

Matter
Names
Formula
Physical properties
Chemical properties
Methods of production
Uses
Comparisons
Predictions

Energy
Form
Production
Measurement
Transformed
Interaction with matter
Examples
Comparisons
Predictions

Process/Procedure
Purpose
Major steps
Occurrence
Relationship to experiment
Comparisons
Predictions

Chapter Correlations
13—Writing Formal Papers
17—Scheduling Student Research

© 2000 by Kendall/Hunt Publishing Company, Cothron, Giese, & Rezba, *Students and Research*.

TABLE 16.10   Writing a Scientific Research Paper

© 1998 PhotoDisc, Inc.

		10 Evaluating for Success	
Name_____ **Period** _____ **Date** _____			
Criteria/Value (100 points)	**Self**	**Peer/ Family**	**Teacher**
Introduction (15)			
Background on IV and DV (5)			
Review of prior research (5)			
Statement of problem (5)			
Methods and Materials (10)			
All materials/Equipment included (5)			
Clear/Precise description (5)			
Results (20)			
Data tables (5)			
Graphs (5)			
Paragraphs of results (5–10)			
Statistical Test—optional (0–5)			
Discussion and Conclusion (20)			
Purpose of experiment (3)			
Major findings (5)			
Support of research hypothesis by data (3)			
Comparison with other research (3)			
Explanations for findings (3)			
Recommendations (3)			
Writing (20)			
Logical organization/Effective transitions (5)			
Sentence/Paragraph structure (5)			
Grammar/Spelling (10)			
General Format (15)			
Abstract (5)			
Title page (2)			
Footnotes (2)			
Bibliography (2)			
Acknowledgments (2)			
Appendix (2)			

Chapter Correlations
13—Writing Formal
      Papers
17—Scheduling Student
      Research

© 2000 by Kendall/Hunt Publishing Company, Cothron, Giese, & Rezba, *Students and Research.*

TABLE 16.11   Presenting Scientific Research

	11 Evaluating for Success		
Name _____ **Period** _____ **Date** _____			
Criteria/Value	**Self**	**Peer/ Family**	**Teacher**
Content (50)			
Background information (10)			
Statement of problem (5)			
Methods and materials (5)			
Results (15)			
Discussion-conclusion (15)			
Questioning (20)			
Knowledge of topic (5)			
Recognition of limitations (5)			
Recommendations for further study (5)			
Acknowledgments (5)			
Presentation of research—use appropriate criteria for presentation.			
Visual display (30)			
Size requirements (6)			
Accurate (6)			
Legible (6)			
Quality photographs/Drawings and so on (6)			
Attractive (6)			
Oral Presentation (30)			
Delivery (15)			
Eye contact (5)			
Volume (5)			
Pace (5)			
Audiovisual Materials (15)			
Relevant to presentation (5)			
Legible/attractive (5)			
Quality slides/transparencies (5)			

© 1998 PhotoDisc, Inc.

© 2000 by Kendall/Hunt Publishing Company, Cothron, Giese, & Rezba, *Students and Research*.

Chapter Correlations
17—Scheduling Student Research
18—Presenting Student Research
19—Preparing to Judge Competitions

Related Web Site

http://www.eduzone.com/
    Tips/science/SHOWTIP2.
    HTM (Simple overall
    checklist)

CHAPTER 17

Scheduling Student Research

© 1998 PhotoDisc, Inc.

Objectives

- Describe a teaching schedule for integrating independent research projects into the curriculum.
- Use a schedule to guide students through an independent research project.
- Develop a student schedule for independent research projects.
- Use a schedule to incorporate independent research projects into the curriculum.
- Develop a teaching schedule for incorporating independent research projects into your existing curriculum.

Ask teachers, "What are the major problems associated with independent student research projects?" and a frequent response will be "insufficient class time." Independent research projects span the academic year, especially if they are to be entered in a science competition. For many students, a science research project is their first independent long-term project. Both time management skills and an increased understanding of the nature of scientific research are important educational outcomes. By modifying laboratory activities already contained in the curriculum, teachers can achieve the dual goals of teaching content and of teaching research skills. Meanwhile, students must learn to simultaneously achieve the multiple goals of learning science content, mastering research skills, and completing their research project. To facilitate student mastery of time management skills, experienced teachers provide both students and parents with a timeline of ac-

tivities. Although competitive deadlines will differ, the timelines provided in this chapter can be used as a guide to assist the teacher, student, and parents with effective time management. The timelines are to be used as guides. Because many unforeseen problems will emerge as students conduct research, assign deadlines but consider them to be flexible, not mandates.

STUDENT SCHEDULE

Few parents had the experience as students of doing research projects. Even fewer students comprehend all that is involved in a research project after just reading the typical science textbook. Begin by teaching the basic concepts of experimental design. When students have mastered the concepts, send a

simple science experiment home. The students can demonstrate their new knowledge and skills while parents learn concepts they probably never encountered in science classes.

Students should research a topic that is of genuine interest to them. Obtain information about student interest through a questionnaire, such as the one shown in Activity 17.1, *What's Your Interest?* Determine the student's favorite school subjects, hobbies, and career goals. Ascertain where each student could conduct a science project—at home, school, local research laboratory, or as part of an after-school job. Inquire about television programs, articles, and books on science topics that are of interest to the student.

Using student interests as the basis, give a motivational presentation on project topics. Ask former students to present their projects, invite local scientists to discuss their research, and view audiovisuals on science projects. Include examples of both non-traditional and traditional research topics.

Give students a timeline for completing research projects (see Table 17.1 *Student Timeline*). Emphasize that each component of the project will be taught in class and that students will be provided with the evaluation criteria. Establish student folders for storage of each component of the project as it is completed.

Ask students to invite their families to a meeting about student research projects. At the meeting, showcase a range of projects from prior years. Include a variety of topics and of student achievement levels. Remember, if only winners are displayed, many parents and students may view the research project as an unachieveable objective. Consider having students demonstrate and teach various parts of experimental design through a quick, novel activity. Provide parents with copies of the student timeline and reiterate the sequential development of the project based on classroom activities.

TEACHING TIMELINE

By modifying activities already in the curriculum, students can be taught the skills needed for research throughout the year. Generally, teachers prefer a timeline that spreads instruction on research skills throughout the year. Students also perceive the research skills as more relevant when immediate applications of the topics to their research project occur. Teaching research topics throughout the school year also allows both the students and the teacher opportunities for ex-

Nontraditional	The Effect of Tractor Noise on Germinating Plants (Agricultural Science)
	A Study of Kerosene Heaters in the Home (Consumer Science)
	The Caloric Value of Imitation Bacon versus Real Bacon (Home Economics/Health)
Traditional	The Effect of Different Rotations on the Color, Height, Germination, and Mass of Annual Rye Grass Seeds (Botany)
	Qualitative and Quantitative Analysis of Sodium Aluminum Phosphate and Tartaric Acid in Pancake Mixes (Chemistry)
	Limb Formation: A Result of Cell Movement or Cell Division? (Developmental Biology)
	Hurricanes and Tornadoes: The Relationship Between Land Temperature and the Strength of Hurricanes (Earth Science)

ACTIVITY 17.1 • What's Your Interest?

1. Where would you like to work on your project?

   home
   school
   job
   local college or research lab

2. Which type of project interests you?

   practical project
   theoretical project

3. What is your favorite school subject?

   science
   mathematics
   fine arts (art, music)
   humanities (language, social studies, psychology)
   health and physical education
   vocational

4. If science is your favorite area, which science is your favorite?

   biology
   chemistry
   physics
   earth
   environmental

5. If biology is a favorite, which area appeals to you?

   animal behavior
   genetics
   botany
   zoology
   biochemistry
   medical science
   developmental biology

6. What is your favorite hobby?

7. Have you read a journal article or book that appealed to you? What was it about?

8. What are your career interests?

© 2000 by Kendall/Hunt Publishing Company, Cothron, Giese, & Rezba, *Students and Research*.

TABLE 17.1　Student Timeline

Assignment	Date Due	Value of Assignment
September		
1. Activity: What's your interest?	September 15	Homework grade
2. Note cards on three popular journal or newspaper articles	September 22	Minor grade *(Rating Sheet 5)*
3. General project topic and action of interest	September 22	Homework grade
4. Note cards on five general sources	September 27	Minor grade *(Rating Sheet 5)*
5. Complete the *Four Question Strategy* for your general project topic	September 29	Minor grade *(Rating Sheet 1, Part II)*
October		
6. Note cards on three scientific articles	October 6	Minor grade *(Rating Sheet 5)*
7. Draft experimental design diagram	October 10	Homework check *(Rating Sheet 1, Part I)*
8. Note cards on five technical manuals and/or community interviews	October 13	Minor grade *(Rating Sheet 5)*
9. Draft list of materials and equipment	October 18	Homework check
10. Parental permission forms for use of live organisms, chemicals, or hazardous procedures	October 23	REQUIRED BEFORE PROCEEDING
11. Draft procedures	October 26	Homework check *(Rating Sheet 2)*
November		
12. Review of the literature	November 9	Major grade *(Rating Sheet 9)*
13. Progress report one or request for revisions/extension	November 16	Homework check *(Rating Sheet 6)*
14. Progress report two or request for revisions/extension	November 30	Homework check *(Rating Sheet 6)*
December		
15. Progress report three or request for revisions/extension	December 8	Homework check *(Rating Sheet 6)*
16. Draft data analysis (tables, graphs, paragraphs)	December 15	Major grade *(Rating Sheet 3 or 7)*
January		
17. Statistical test (if appropriate) and prepare draft table of results and/or paragraph	January 5	Minor grade *(Rating Sheet 8)*
18. Draft conclusion	January 10	Minor grade *(Rating Sheet 7 or 8)*
19. Draft research paper	January 30	Major grade *(Rating Sheet 4 or 9)*
February		
20. Final research paper	February 10	Major grade *(Rating Sheet 4 or 10)*
21. Revise/edit research paper and place in format for competition	Dependent	
March–May		
22. Prepare visual display or oral presentation	March 3	Major grade *(Rating Sheet 11)*
23. Participate in competitive events		
District Science Fair	March 13–16	
Regional Science Fair	April 2–4	
State Academy of Science	May 25–28	
24. Attend Science Night	May 15	

© 2000 by Kendall/Hunt Publishing Company, Cothron, Giese, & Rezba, *Students and Research*.

tensive practice with each component before proceeding to the next phase. Research activities fall into six general phases:

September: What is a research project? **Basic concepts of experimental design are introduced, motivational project ideas presented, and expectations clarified for students and parents.**

September to October: What is my project? **Through brainstorming activities and library research, project topics are refined, materials and equipment identified, and procedures developed for conducting research. Proposed experiments are summarized through reviewing the literature and finalizing experimental designs.**

November to January: How do I collect and analyze data? **Students refine techniques; collect and analyze data through tables, graphs, and statistical tests; and summarize results.**

January to February: How do I write about scientific research? **Components are edited, revised, and integrated into scientific research reports. Reports are modified to the specifications of the competitive events the students enter.**

March to May: How do I present research? **Students learn how to make visual or oral presentations. Presentations are modified to the specifications of the competitive events the students enter.**

May to June: I did it! **Student achievements are showcased; parent, school, and community support is acknowledged and publicized.**

The following teaching timeline includes concepts to be formally addressed in the classroom, student responsibilities outside class, and suggested evaluation techniques (see Table 17.2 *Teaching Timeline*). Classroom activities are correlated with basic and advanced concepts of experimental design included in this text. Evaluation techniques are based on rating sheets included in Chapter 16. Parent letters are included in Chapter 14.

MINIPROJECTS

Miniprojects in class provide an alternative to independent research projects. Using classroom materials, students can work as individuals or groups to design and execute a simple project in class. After learning the basic concepts of experimental design and the *Four Question Strategy*, students decide on a potential topic. Topic selection is restricted by classroom materials and prompts. General library skills are taught and practiced as students write an introduction. After direct instruction, students write and revise procedures for their investigation. Techniques for collecting and analyzing data that are appropriate for the students' ability are applied to the project. Structured models for writing about results and conclusions are modeled and implemented by students, with the writing culminating in a report appropriate for the students' ability. The miniproject culminates with short presentations in class.

© 1998 PhotoDisc, Inc.

Miniprojects in class can be an excellent alternative for middle and high school students with limited home support. By conducting several miniprojects during the year, students have opportunities for guided practice and the refinement of their research skills. Miniprojects in class can also provide a useful review for students who were taught research concepts in previous science classes. The model can be differentiated to provide for students of varied ages and abilities and for individual or small group work. Table 17.3 *Miniproject Schedule* suggests a sequence for conducting miniprojects.

Related Web Site

http://www.eduzone.com/Tips/
science/SHOWTIP2.HTM
(12 week timetable)

TABLE 17.2   Teaching Timeline

Major Activities	Concepts/Activities Formally Addressed in Class	Student Responsibilities Outside Class
What is a Research Project? (September)		
Designing experiments (Chapters 1, 2, 12)	Teach and practice basic concepts of experimental design	Complete homework assignments on scenarios
Parent letter 1: Family involvement in a simple science experiment (Chapter 14)	Distribute letter	Deliver letter and invite family
Potential project topics	Motivational presentation on interesting project topics	Complete activity: What's your interest? (Homework Check)
Parent letter 2: General overview of research projects and invitation to a meeting (Chapter 14)	Distribute letters	Deliver letter and encourage participation
Timeline for completing research project (Chapter 17)	Distribute timeline	Follow timeline
Establish student folders (Chapter 17)	Expectations for students' project folders	Maintain project folders
Parent meeting	Remind students	Attend meeting and bring parents
What is My Project? (September–October)		
Library skills: Establish an interest—topic connection (Chapter 7)	Discuss popular journals and newspapers; teach skills of referencing and taking notes	Complete note cards on popular journal or newspaper articles (Rating Sheet 5—minor grade)
Student/parent letter 6: Insufficient progress on project topic (Chapter 14)	Distribute letters to appropriate students/parents	Deliver letter to parents make appointment with teacher
Generating ideas for project (Chapter 3)	Teach the Four Question Strategy Practice the Four Question Strategy using a variety of props	Complete assignments on the Four Question Strategy (Rating Sheet 1, Part II—homework check)
Parent letter 8: Library research (Chapter 14)	Distribute letters	Deliver letter
Library skills: Use general references to narrow topic (Chapter 7)	Teach library classification systems, card catalog, references, scanning, note-taking	Complete note cards on general sources (Rating Sheet 5—minor grade)
		Complete Four Question Strategy for project topic (Rating Sheet 1, Part II—minor grade)
Library skills: Use of scientific journals to clarify variables (Chapter 7)	Teach use of scientific indices, referencing, scanning, note-taking	Complete note cards on scientific articles (Rating Sheet 5—minor grade)
		Complete draft experimental design diagram for project (Rating Sheet 1, Part I—homework check)
Writing procedures (Chapter 4)	Teach and practice writing procedures	Complete assignments on procedures (Rating Sheet 2—homework check)

(continued on the following page)

TABLE 17.2 (continued)

Major Activities	Concepts/Activities Formally Addressed in Class	Student Responsibilities Outside Class
Library skills: Consult technical handbooks, manuals, and community agencies to refine procedures *(Chapter 7)*	Discuss technical manuals, handbooks, and community agencies; teach interviewing skills	Complete note cards on technical materials and community interviews *(Rating Sheet 5—minor grade)*
Review of materials and equipment list *(Chapter 4)*	Discuss safety considerations, humane treatment of organisms, funding, community resources	Submit draft list of materials and equipment for project *(homework check)*
Parent letters 3–5: Permission for use of organisms, chemicals, hazardous procedures *(Chapter 14)*	Distribute letters to appropriate students	Deliver letter and return signed permission; complete draft procedures for project *(Rating Sheet 2—homework check)*
Write review of literature *(Chapter 13)*	Provide structured outlines/questions for students to use in reviewing adequacy of library research; teach requirements for introduction	Conduct additional library research as needed
		Write review of the literature *(Rating Sheet 9—major grade)*

How Do I Collect and Analyze Data? (November–January)

Major Activities	Concepts/Activities Formally Addressed in Class	Student Responsibilities Outside Class
Conduct research	© 1998 PhotoDisc, Inc.	Submit progress report one or request for deadline extensions *(Rating Sheet 6—homework check)*
		Submit progress report two or request for revisions and deadline extensions *(Rating Sheet 6—homework check)*
Student/parent letter 7: Insufficient progress on research *(Chapter 14)*	Distribute letters to appropriate students/parents	Deliver letter to parents
		Make appointment with teacher
		Submit progress report three or request for revisions and deadline extensions *(Rating Sheet 6—homework check)*
Non-inferential statistical techniques for analyzing data and writing results *(Chapters 6, 8, 9, 10)*	Teach appropriate data analysis and writing skills to your students • Simple data table and graphs • Quantitative and qualitative data tables	Complete homework assignments on analyzing data and writing results
		Prepare draft data analysis: Tables, graphs, paragraphs *(Rating Sheets 3 or 7—major grade)*
Inferential statistical techniques for analyzing data *(Chapter 11)*	Teach statistical tests • *t* test • chi-square	Complete homework assignments on statistical tests
		Conduct statistical tests (if appropriate) and prepare draft tables of results and/or paragraphs *(Rating Sheet 8—minor grade)*
Writing a conclusion *(Chapters 6, 9)*	Teach how to write a conclusion	Complete draft conclusion for project *(Rating Sheet 7 or 8—minor grade)*

(continued on the following page)

TABLE 17.2 (continued)

Major Activities	Concepts/Activities Formally Addressed in Class	Student Responsibilities Outside Class
How Do I Write About Scientific Research? (January–February)		
Research paper *(Chapter 13)*	Review components of research paper and criteria for evaluation	Complete draft research paper *(Rating Sheet 4—major grade)*
Parent letter 9: Review of draft research paper *(Chapter 14)*	Distribute letters	Deliver letter
		Prepare final research paper *(Rating Sheet 4 or 10—major grade)*
Parent letters 10–11: Participation in competitive events *(Chapter 14)*	Distribute letters (if appropriate)	Deliver letter (if appropriate)
Preparing written papers for competition *(Chapter 13)*	Revise competitive requirements, distribute forms, and so on	Revise/edit research papers and place in appropriate format for submitting for competition
How Do I Present Research? (March–May)		
Preparing oral and visual displays of projects *(Chapter 18)*	Discuss components of good oral and visual displays and tips for preparing	Prepare oral or visual displays of project *(Rating Sheet 11—major grade)*
Assisting students involved in competition *(Chapter 18)*		Fulfill requirements and meet with teacher at appointed time
Publicity on students involved in competition		
Parent letter 12: Requirements for upcoming competitive events *(Chapter 14)*	Distribute letters	Deliver letter and obtain permission to participate
Competitive events *(Chapters 18, 19)*		Student participation in events
I Did It! (May–June)		
Parent letter 13: Letter of appreciation and invitation to attend science night *(Chapter 14)*	Distribute letters	Deliver letter and encourage parents to attend
Science night: Showcase achievements of *all* students		Attend science night with parents

© 2000 by Kendall/Hunt Publishing Company, Cothron, Giese, & Rezba, *Students and Research*.

TABLE 17.3   Miniproject Schedule

Major Concepts	Classroom Activities (Group or Individual Work)
What Is a Resource Project?	
Developing basic concepts *(Chapters 1, 2)*	Teach & practice basic concepts of experimental design *(Rating Sheet 1, Part I)*
What Is My Project?	
Generating ideas for projects *(Chapter 3)*	• Teach & practice the *Four Question Strategy* • Generate ideas for investigations using designated prompts • Group decision on potential topic to investigate
Using library references to narrow topic *(Chapter 7)*	• Teach & practice appropriate library skills to students using textbooks & school library materials • Group completes note cards *(Rating Sheet 5)*
Writing an introduction *(Chapter 6 or 13)*	• Teach & practice writing an introduction for a simple report *or* scientific research paper • Group prepares introduction for an investigation
Preparing experimental design *(Chapters 1, 2, or 12)*	• Group prepares experimental design diagram *(Rating Sheet 1, Part I)*
Writing procedures *(Chapter 4)*	• Teach & practice writing procedures • Group prepares procedures for investigation *(Rating Sheet 2)*
How Do I Collect and Analyze Data?	
Constructing data tables *(Chapter 5)*	• Teach how to make a simple data table • Group prepares data table for investigation
Conducting an investigation	• Group conducts investigation & records data *(Rating Sheet 3, Part I)*
Analyzing data & writing results *(Chapters 5, 6 or 8, 9, 10, 11)*	• Teach appropriate data analysis and writing skills for students—simple data tables/graphs, descriptive statistics, inferential statistics • Group prepares data analysis for investigation—tables, graphs, paragraphs *(Rating Sheet 3, 7 or 8)*
How Do I Write About Scientific Research?	
Write a conclusion *(Chapters 6 or 9, 10, 11, 12, 13)*	• Teach how to write a conclusion • Group prepares conclusion *(Rating Sheet 3, 7, or 8)*
Reporting scientific research *(Chapter 6 or 13)*	• Teach appropriate type of report for students—simple report or scientific research paper • Group prepares written report *(Rating Sheet 4 or 10)*
How Do I Present Research?	
Presenting scientific research *(Chapter 18)*	• Teach how to make an oral or visual presentation • Group prepares presentation *(Rating Sheet 11)* • Group presents scientific research
I Did It!	

© 2000 by Kendall/Hunt Publishing Company, Cothron, Giese, & Rezba, *Students and Research.*

PART FOUR

Strategies for Successful Science Competitions

✓ Explore options for maximizing student success and attitudes about scientific inquiry:

- make effective presentations
- design attractive displays
- prepare students for success
- train inexperienced judges
- select the right competition

··········· C H A P T E R 18 ···········

Presenting Student Research

Objectives

- Identify important features of good oral presentations.
- Make an oral presentation on an independent research project.
- Identify important features of good displays for science fairs and poster sessions.
- Make a science fair or poster display on an independent research project.
- Prepare students to make an oral presentation or display on an independent research project.
- Enhance students' presentation skills by incorporating a variety of oral presentations and visual displays into the curriculum.

National Standards Connections

- Communicate and defend a scientific argument (NSES).
- Think critically and logically to make the relationships between evidence and explanations (NSES).

In contrast to written scientific papers, which follow a standard format, oral and visual presentations of scientific research exemplify diversity. Students are frequently asked to present research through oral reports, poster sessions, and displays. The method of presentation differs with each competitive event. Science fairs, the dominant competitive event at local and regional levels, typically require a three-panel display board. Poster sessions held by the American Academy of Science provide a low-cost alternative to science fair displays and are particularly appropriate for younger and disadvantaged students. Competitive events sponsored by industries, special interest societies, academies of science, and government agencies require oral presentations of projects previously selected from written reports. Typically, competitive events that require displays or posters also require written reports and oral presentations of projects to judges. This chapter discusses the general format for each type of presentation. Because there are many excellent resources related to preparing and conducting science fairs, major emphasis will be given to the oral component of a presentation. Strategies for incorporating practice of the oral and visual aspects of presentations into daily science instruction are also provided.

SCIENCE FAIR DISPLAYS

Most local science fairs follow the requirements of the International Science and Engineering Fair for exhibit size and components. Exhibits cannot exceed a depth of 76 cm (30 in), width of 122 cm (48 in) and height of 274 cm (9 ft) including the height of a table. Guidelines prohibit displays of vertebrate and invertebrate animals; photographs of surgical techniques such as autopsies; exhibition of most human parts; Class III or IV lasers; and fuels, foods, microbes, and chemicals that endanger public safety. Because of state quarantine requirements, displays of plant or agricultural products are discouraged. Consult the rules of the International Science and Engineering Fair, published yearly, for specific requirements. The format of a science fair display board parallels the components of a simple science report: title, statement of problem, procedure (methods-materials), results, and conclusion (see Chapter 6). Figure 18.1 depicts the traditional position of these components on a science fair display board.

Title

A title may state the specific independent and dependent variables being investigated or may be worded creatively to capture the readers' interest.

- The Effect of Bay Leaves and Cucumbers on Cricket Behavior
- Repel Crickets with Cucumbers and Bay Leaves
- Methods of Warfare on Crickets

Statement of the Problem

The essentials of the research must be clearly communicated through the problem statement. On display boards, the problem is frequently stated as a question followed by the specific hypotheses to be tested.

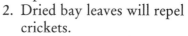

Problem: Will cucumber skins and bay leaves repel crickets?

Hypotheses: 1. Dried cucumber skins will repel crickets.
 2. Dried bay leaves will repel crickets.
3. A combination of dried cucumber skins and bay leaves will be most effective in repelling crickets.

Procedure

The procedure (methods-materials) for the experiment may be communicated as lists or written in paragraph form. Because of limited space, the procedure must be stated succinctly. Consult science fair guidelines for requirements. If an option exists, use the most appropriate format for your students' age and experience (see Chapters 4, 6, 13).

One hundred grams of bran were placed in one corner of a 20-gallon aquarium. Another 100 g of bran ringed with 25 g of dried bay leaves were placed in the opposite corner. Fifty crickets were placed in the aquarium. The distribution of crickets and the mass of bran consumed after 24 hr were recorded.

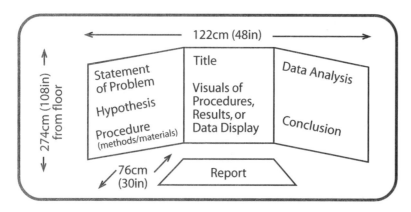

Figure 18.1 General Diagram of Science Fair Display.

Allowing a 1-day recovery period between each trial, the procedure was repeated 4 times. Similarly, the response of crickets to two piles of bran alone (control), to cucumber skins surrounding one pile of bran, and to a combination of bay leaves and cucumber skins surrounding another pile of bran were determined.

Results

Sufficient data tables and graphs must be included to communicate the findings and to show the extent to which the data supports the research hypotheses. Typically, space does not permit the inclusion of numerous tables, graphs, or a lengthy discussion of the results and conclusions. Photographs or diagrams of experimental results are particularly effective. For example, the distribution of crickets could be depicted through photographs or diagrams, whereas the mass of bran consumed could be displayed through a graph. Brief sentences summarizing the data could accompany the diagrams and graph (see Figure 18.2). Students must determine which display techniques best communicate their experiment.

Supplementary data tables and graphs should be placed in the written report. Remind students to refer to them when making an oral presentation to the judges.

Conclusion

In the conclusion section of a science fair display, the student summarizes major findings and the extent to which the results support the research hypotheses. Findings must be written concisely in paragraph form or as a list. A brief explanation of findings is also appropriate as part of the conclusion. Display space will not permit a lengthy discussion of results as in a written report. Major recommendations for additional research and improvements may be cited if space permits.

Effective warfare against crickets can be waged with bay leaves but not with dried cucumber skins. Research data supported the hypothesized effectiveness of bay leaves but NOT the hypothesized effectiveness of cucumber skins NOR the superior

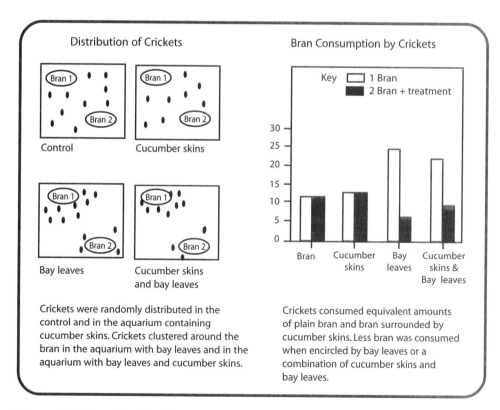

Figure 18.2 Data Displays for Science Fair Display, Methods of Warfare on Crickets.

repelling power of the combination. Future experiments should address the effectiveness of fresh materials and of mixing food and materials. Different cricket samples should be used to eliminate potential saturation or adaptation.

Constructing Display Boards

Numerous methods exist for constructing display boards. Appropriate materials include plywood, plexiglass, reinforced poster board, corrugated cardboard, and styrofoam sheets. The three-panel board may be hinged or reinforced with tape. Inexpensive project display boards are available in many large office supply stores. All lettering should be neat and legible; spelling should be carefully checked. Consider artistic appeal by selecting complementary colors that enhance photographs and display materials. Excellent publications on photography in science research are published by the Eastman Kodak Company. Plan the display carefully. Remember, it is the advertisement for the project. A sloppy display will dissuade judges and the public. Strategies for assessing visual displays are provided in Chapter 16, *Assessing with Rating Sheets.*

POSTER DISPLAYS

Poster displays, accompanied by written reports, provide a low-cost alternative to science fair displays. For younger or economically disadvantaged students, posters can comprise the local science fair. With older students, valuable experience can be provided by requiring these succinct displays of data. Brief oral reports, 3 to 5 minutes long, can accompany the poster. A poster display includes a title, an abstract of the experiment, and critical tables and graphs to communicate the findings. A simple report or a formal research paper may be stapled to the lower edge of the poster. Posters that are displayed in classrooms or along school halls maximize opportunities for students and parents to view student research projects. A sample poster is shown in Figure 18.3. Strategies for assessing visual displays are provided in Chapter 16, *Assessing with Rating Sheets.* Because all students explain their research to judges, other students, and community members, good communication skills are essential. These skills can be enhanced by applying the strategies for formal oral presentations described in the following sections.

ORAL PRESENTATIONS

Typical oral presentations required for many competitive events span approximately 10 minutes and parallel the components of the formal research paper. Frequently, 5 minutes of questioning by judges follows the oral presentation. When the format of a competition makes a display board or poster inappropriate, other visual aids, such as transparencies and slides, are critical to an effective presentation. Major components of the presentation and representative types of visual aids are outlined in Table 18.1 *Suggested Sequence and Visuals for Oral Presentations.*

Students will be more comfortable with the prospect of public speaking if they know and practice what will be expected. Provide students with copies of Table 18.1 or a simple list of steps.

- Begin by telling the audience about yourself and how you became interested in your project. Include your name, grade, and school.
- Describe your problem and give important background information on the variables in your study and in related research.
- State your purpose and research hypotheses.
- Describe the procedure you followed to test the hypotheses. Be sure to include information on your independent and dependent variables, the constants, the control, and the number of repeated trials. It is important that you also acknowledge any help you received in conducting the experiment.
- Explain your results using transparencies or slides of tables and graphs.
- Share your conclusions and state the extent to which they supported the original hypothesis.
- Suggest areas for future study and for improvement of the experiment.

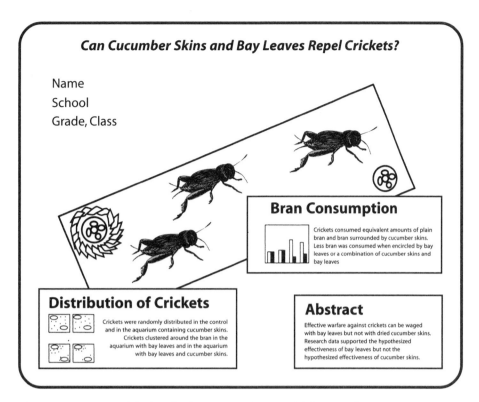

Figure 18.3 Poster Display for Science Project on Cricket Warfare.

Students can transform their written report into an oral presentation by using the following steps.

- Underline the most important facts in the research paper. Note the best graphs and tables. Describe visuals that may need to be added.
- Read the paper to a teacher, parent, or another student. Describe the visuals you will use. Ask for feedback on posture, speech, volume, inflection, speed, and proposed visuals.
- Write the important information on note cards. With a colored pen, designate visuals to be used.
- Prepare visuals.
- Using the visuals, read the note cards to a teacher, parent, or fellow student. Record the reading. Obtain feedback.
- Practice, practice, practice—in front of the mirror, at home, or with other students.
- Practice the presentation in front of a class. Obtain feedback.
- Continue to polish the presentation until you can speak (not read) from the note cards and can effectively integrate the visuals.

Finally, remind students that it is normal to be nervous, especially when they begin. Tell them to take a deep breath, stand straight, and begin. Discuss the importance of creating a good first impression by looking your best, by not chewing gum, and by maintaining eye contact with the judges and the audience. Encourage students to be friendly and enthusiastic about their projects. Enthusiasm is contagious. It can motivate the audience to be more attentive. Review basic speaking principles including the importance of speaking clearly, at a reasonable pace, and loudly enough to be heard by everyone. Remind students to stand to the side of visual aids and to point to the appropriate information on the visual aids. Strategies for assessing students' oral presentation skills are provided in Chapter 16, *Assessing with Rating Sheets.*

Practicing Oral Presentations

The prospect of making an oral presentation is very threatening to students. Most students will need frequent nonthreatening, low-risk opportunities to develop their oral skills and to overcome

TABLE 18.1 Suggested Sequence and Visuals for Oral Presentations

Presentation Components	Representative Types of Visual Aids
Introduction	
1. Introduce yourself. 2. Tell your audience how you became interested in the topic. 3. State the problem. 4. Review pertinent background information on variables and prior research. 5. State the research hypotheses.	 • Picture, diagram, table, or graph that illustrates the problem and sets the stage for the presentation • Photographs or diagrams of experimental subjects or phenomena • Table summarizing similarities and differences among research studies • List of research hypotheses
Procedure (Methods-Materials)	
1. Describe the design of the experiment. 2. Describe the experimental procedures (methods and materials). 3. Explain how the data were analyzed.	• Diagram of the experimental design • List of materials • Diagrams or photographs of special apparatus • List of steps or a flow diagram • Photographs of experimental stages
Results	
1. Display the results. 2. Describe the results.	• Tables and graphs for each type of data • Photographs of treatment groups
Discussion-Conclusion	
1. Summarize the major findings, including support for the research hypotheses. 2. Compare findings with prior research. 3. Suggest improvements, topics for further study, and potential applications.	• List of major findings • List of improvements, future research topics, and applications

© 2000 by Kendall/Hunt Publishing Company, Cothron, Giese, & Rezba, *Students and Research*.

their fears of speaking in front of others. Effective strategies for developing students' presentation skills include practicing one component at a time, modeling of research presentations, and speaking in a variety of environments.

Individual Components

Early in the academic year, devote a few minutes of class time for several days to practicing oral presentation skills. Begin by having students describe the purpose of their research projects in a few sentences while standing at their desk. The comfort provided by the security of the desk can help offset  the fear of speaking while standing. Speaking to the whole class may be too intimidating for some students. If so, arrange the students in groups and conduct the class as if the groups were concurrent sessions at a conference.

Later, ask students to stand at their desks and to give a 2-minute overview of their research project that includes the purpose, rationale, reason for interest, and hypotheses. During these sessions, both the desk and the structured format provide a sense of security.

When students have finalized the experimental designs for their research projects, have them stand at a lectern and describe their designs. The description should include the independent and dependent variables, hypotheses, constants, control, and repeated trials. Provide practice with audiovisuals by having students use the experimental design diagram on a transparency, newsprint, or the blackboard as the focal point of their presentation. Because the students are away from their desks, the audiovisual can provide security. Allow 5 minutes for each presentation, including questioning by students. Again, several small groups of students can function at the same time. Similarly, students can present procedures, results, and conclusions of their research studies.

Modeling

Because few students have seen scientists present research, modeling of research presentations is essential. Ask former students to make guest presentations of their research or show videotapes of student presentations. Follow the presentations with discussions of the major components, types of audiovisuals, and presentation styles. As an alternative strategy, the teacher can present a research paper. By stopping at various points, the teacher can illustrate and emphasize important components of an oral presentation.

Students will feel more comfortable about speaking if they understand the atmosphere of the presentation. Ask former students to describe the setting including the type of room and the people in the audience—the judges, other presenters, students, teachers, and parent observers. Judges of science competitions can also play an important role. Use the program from a prior competitive event to identify judges to visit your class. The judges can describe student presentations from their point of view—what most students do, what students should do, what criteria judges use, what questions judges most frequently ask of students, and so on. Some teachers use judges in a different way. They ask two to three students to present their proposed research projects to an invited judge. The judge asks questions about the presentations including the introduction, purpose, method-materials, results, and conclusions. Remaining students form pairs; one presents and the other uses the visiting judge's comments and questions as a model for quizzing the partner. This strategy has the advantage of letting students know in advance the types of questions judges might ask. A potential disadvantage is that a public critique of proposed projects by a judge can be very hard on sensitive students. Select your judge and presenters with care.

Varied Environments

Sometimes there is no substitute for the experience a student can gain from speaking to a whole class of peers. This is especially true for students who will make formal presentations at competitive science events. Arrangements should be made for them to speak before an entire class if at all possible. This could be their own science class, another teacher's class, or even a class of younger students. If developing oral presentation skills can be coordinated with the English department, students can speak to English classes as well. Students can provide valuable feedback on delivery skills.

Family involvement may offer several interesting practice opportunities for students. Students can ask their family to serve as their audience. Even speaking to just one parent or sibling can be good practice. Students can also model the techniques used by many accomplished speakers including speaking in front of a mirror (provides information about facial expression and nonverbal communication skills in general) and speaking into a tape or videotape recorder (gives information about many qualities including tone, pitch, and inflection). Some students have reported that they tried the humorous technique of using the family pet as a captive audience; however, the pet often yawned, fell asleep, or just wandered off!

When class time is just not available, special afternoon and evening practice sessions can be conducted. Parents, other family members, and friends can be encouraged to attend these sessions to serve as an audience for members of the science class (see Chapter 14, *Encouraging Parental Support*). After school sessions give students the opportunity to practice their presentations in front of supportive friends and parents, a much less threatening experience than in front of strangers, particularly judges.

USING TECHNOLOGY

Encourage students to use a presentation graphics program to describe their experiment and to present their findings. Provide opportunities for students to learn a program such as PowerPoint to create slides, outlines, and tables to share what they know. Visuals provide a framework for presentations and may reduce student anxiety about public speaking. In helping students prepare for any competitive event, be sure to check the rules regarding presentation. Caution students to have a back up plan if a problem arises with equipment.

Related Web Sites

http://ibms50.scri.fsu.edu/~dennisl/
  CMS.html
http://134.121.112.29/sciforum/
  writing.html (Scientific Poster)
http://www.eduzone.com/Tips/science/
  showtip4.htm (Display Boards)
http://www.eduzone.com/Tips/
  science/SHOWTIP2.
  HTM (logistics for
  organizing science fair
  competition)

Preparing to Judge Competitions

© 1998 PhotoDisc, Inc.

Objectives

- Justify preparing judges for competitive events.
- Describe essential information to communicate to judges about student research projects.
- Describe ways to maximize opportunities for students to learn more about research by interacting with judges.
- Develop procedures for preparing judges for science fair competitions.

Becoming a parent and judging a science competition have something in common. In both cases, the need for special skills is realized only after one is fully committed. Just as most parents would benefit from some training in parenting, most judges would benefit from training in the skills needed for judging science competitions. This is especially true for judges of middle school students or secondary students entering their first competition.

One must be a realist, however. It is already difficult to locate sufficient numbers of scientifically literate people who are willing to donate time to be a judge. To find people who would also be willing to attend a training session would make the task nearly impossible. Thus, the following **not so tongue-in-cheek letter** is presented as a model letter to be sent to prospective judges for science competitions. It is intended to focus a judge's attention on both the development of the student researcher and on the critical balance of process and content in science projects.

DEAR DR. B. EXEMPLARY JUDGE:

Congratulations and many thanks for accepting the role of a judge at our upcoming science competition. We appreciate your time and expertise. You were identified as a scientist or as a professional in a science-related career, but more importantly, you were selected because of your interest in the education of our youth. As chief judges, we look forward to meeting with you briefly on the morning of the competition. At that time we will make a few general announcements, reassign judges as necessary, and answer any last-minute questions that may arise. In the interim, we wish to share some thoughts about the important role you are assuming.

As a judge you will be a larger-than-life adult. The younger or less experienced the student, the larger you will appear to be. You will be wearing a badge with **JUDGE** on it, mainly as a free pass to lunch and to distinguish you from the parents, but the students will not know that. To the student you are super smart, a college graduate, and a **SCIENTIST** of one sort or another. To many, if not most, you will be very intimidating. Their experience with you can result in emotional as well as intellectual growth. From a student's perspective, your task is to decide who gets the prized tokens of success— the awards, trophies, and ribbons. Yet, science competitions are best viewed not as a final evaluation but as a learning experience in the pursuit of excellence.

Critical Project Content

Students need much encouragement to continue perfecting their skills in designing and analyzing experiments. A sample evaluation form on which judges can rate each student's project is enclosed with this letter (see Table 19.1 *Science Project Evaluation Form*). The instrument is primarily designed to provide feedback to the student on mastery of scientific principles, components of experimental design, and data analysis. Using a 3-point scale, you will rate the student in seven general areas:

Background Knowledge: Knowledge of basic scientific concepts related to the experimental topic; advanced projects may include a formal literature review;

Experimental Design: Articulation of hypothesis, independent and dependent variables, constants, control, and repeated trials;

Procedure (Materials-Methods): Clear and precise description of materials used and steps followed;

Results: Presentation of data in tables, graphs, and summary paragraphs;

Conclusion: Clear, concise statement of the major findings, an interpretation of data, and suggestions for further research;

Display: Attractive, accurate, legible display consistent with science fair regulations;

Interview: Student's ability to communicate the scientific basis of the experiment, to explain principles of experimental design and data analysis, and to relate their experiment to their world.

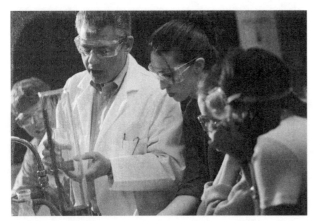

At a minimum, it is important for the student to identify the independent and dependent variables, the hypothesis, the control, the constants, and the repeated trials of the experiment. Each student should also be able to tell you the rationale for each component. Do not assume that because a particular component (such as a control) is part of the experiment, the student understands the concept or can describe why it is an important part of the experiment.

TABLE 19.1 Science Project Evaluation Form

Name _____ Number _____

Please circle: 3 points for very strong; 2 points for moderately strong; 1 point for weak or nonexistent.

Criteria	Point Value		
Background Knowledge			
Key scientific concepts	3	2	1
Literature review	3	2	1
Experimental Design			
Hypothesis: Testable relationship between variables	3	2	1
Independent Variable (IV): Factor purposefully changed	3	2	1
Dependent Variable (DV): Factor that responds	3	2	1
Constants: Factors kept the same	3	2	1
Control: Used as a standard for comparison	3	2	1
Repeated Trials: Number of subjects or times repeated	3	2	1
Procedure (Materials-Methods)			
All materials included	3	2	1
Clear, precise procedure	3	2	1
Results			
Data Tables: IV, DV, derived quantities, units	3	2	1
Graphs: Correct type, scales, title, line-of-best-fit	3	2	1
Summary: Sentences/Paragraphs about tables/Graphs; statement of how data supports hypothesis	3	2	1
Conclusion			
Major findings	3	2	1
Interpretation	3	2	1
Suggestions for further study	3	2	1
Social significance/Application	3	2	1
Display			
Attractive and legible	3	2	1
Accurate	3	2	1
Consistent with fair regulations	3	2	1
Interview			
Communicates scientific basis	3	2	1
Describes design principles	3	2	1
Explains data analysis	3	2	1
Recognizes limitations	3	2	1

© 2000 by Kendall/Hunt Publishing Company, Cothron, Giese, & Rezba, *Students and Research.*

Look for confounding variables, such as using personal friends or pets as subjects. Also, watch for students using themselves as subjects in determining the significance of differences between experimental and control groups. Some students have not comprehended the unintentional bias introduced by an experimenter serving as a subject. Others may not have an alternative other than themselves. Questioning a student about unintentional bias may be interpreted as a reflection or attack on their honesty. Probe but do not press for maturity that is not there.

Another very common design problem is the use of too few subjects in the experimental and control groups. Leading questions help. "Would using six plants rather than one or two for each amount of fertilizer have strengthened, weakened, or not affected your experiment? Why?" Students also need to understand that negative results are just as valid and important as positive findings. Data that do not support a research hypothesis are as meaningful as data supporting it.

As a scientifically literate adult, you frequently think in terms of multiple variables. Such a feat is beyond the maturity level of many students. Some young students may well be limited to reasoning that involves the interaction of one independent and one dependent variable. They may not have the intellect to simultaneously consider the effects of other variables on their project. Again, probe for understanding but do not push the student.

In applying the criteria, always consider the students' maturity. For example, a middle school student's review of the literature may be limited to the citing of a middle school text or two, and in exceptional cases a magazine or newspaper article. The average high school student, however, could be expected to use literature indexes and some journal articles; and a superior high school project should have as extensive a literature review as was permitted by local library resources. Similarly, an adequate data analysis for middle school students is finding and comparing groups on the basis of descriptive statistics, whereas a sophisticated high school student's data analysis may include both descriptive and inferential statistics.

Interviewing Students

In some competitions, the students are not present when the judges are reviewing their projects. Judges must base their decisions solely on the project display. In other events, judges are sent copies of student research papers in advance. Judges are advised to read and preliminarily evaluate the research but to reserve final judgment until after the student interview and presentation. In this competition, students will be present. You will have both the benefit and responsibility of talking with each student experimenter in a section.

Before you begin the interviews, determine how much time you have for each student and stay within the schedule. Each student interview should be approximately the same length and conducted with the same enthusiasm. The weaker the student project, the greater the need for help and supportive suggestions. If students with weak projects are given short interviews, they attribute their lack of success to the interview, not the research project. With a longer interview, they may acquire greater understanding of experimental flaws. We suggest that you and another judge work as a team. By alternating questions, you and another judge will have the necessary time to reflect on each of the student's responses. Team judging also provides time to formulate follow-up questions. Suggested questions for judges are provided in Table 19.2 *Suggested Questions for Judges.*

When you question them, students should be given plenty of time to think. If they appear confused after a pause of a few seconds, rephrase your question. Answers that for you, an experienced scientist, involve simple recall may require in-depth analysis and synthesis of information by students. Be patient and resist any urge to dazzle them. Relax; your best questions may elicit long-term thought but not an immediate response. In our years as judges, many students have approached us later and said, "I was thinking about your questions, and . . ."

TABLE 19.2   Suggested Questions for Judges

Background Knowledge
Why did you decide on this topic?
What is the purpose of your project?
What library information did you find that was helpful?

Experimental Design
What was your hypothesis?
What variable did you intentionally change?
What response did you observe or measure?
What did you intentionally keep the same?
What group did you compare the others against? Why?
How many times did you repeat the experiment?

Materials and Methods
What materials did you use?
What steps did you follow in conducting the experiment?
If you had a mentor, in what ways did the mentor assist you?

Results-Conclusion
What results did you find?
How did your results relate to your original hypothesis?
What conclusion did you make?
If you conducted the experiment again, what would you do differently?
What additional experiments would you suggest?
Which groups in the community would be interested in your experiment?
What recommendations would you make to these groups?
What was the most important thing you learned from the experiment?

© 2000 by Kendall/Hunt Publishing Company, Cothron, Giese, & Rezba, *Students and Research*.

Students should be able to describe their work in detail but not to provide a textbook definition for all associated terms. If you think the term is central to the student's work, probe a bit but be ready for examples rather than definitions. Many students will respond with statements like, "Oh, you mean like when . . ." instead of formal definitions or explanations.

For many students this will be their first experience as a presenter and an important lesson in public speaking. Students will have the added pressure of performing in front of friends, classmates, and new acquaintances, most of whom they perceive as having better projects. Prior competitive experience is only a small source of reassurance. For students, pressure to succeed and fear of failure increase with each competitive level.

The student presenters can and will learn from you, from their successes, and from their mistakes. However, they must comprehend that even if their experimental work does contain flaws, perhaps even fatal flaws, it also has some scientifically redeeming merit. The best judges challenge students to meet high standards with all the rigor that each student can take. Watch students' lower lips and eyes. Look for signs that it would be appropriate to back off. Probe again later if time and student temperament permit. Please end each interview with a smile and an encouraging compliment on some meritorious aspect of the student's experiment.

General Tips

In the event you find a project or two in your section that would be more appropriate in another category, please contact us at once. Because of the broad interdisciplinary nature of students' topics, some projects may be incorrectly placed. We may ask judges from another section to judge that project. Or we may ask you to judge it on the basis of its scientific merits and not on the appropriateness of its topic to your category.

Watch your notes as you move around. Some students will stand on their heads to read them. Use the confidentiality of the judges' conference room to dissect, critically analyze, and rank order student projects. School halls have very big ears.

High standards relative to the students' level of sophistication are important to this competition. However, even the best project in an area may have serious problems. Does it still get first place? World records are not set in every race, yet winners are announced. Imagine being a student and knowing that there were only two entries in the category and yet you received a third place! Such a put down discourages participation; it does not encourage excellence.

Affluence provides opportunities, and money buys expensive materials to be used in projects and displays. Judge student work on content and clarity of design, not the costs. The student who ingeniously used inexpensive materials should be rewarded for good design, not poverty, whereas the more affluent student should not be penalized because of being spared the need to improvise.

A perennial problem is deciding how much of each project is the student's work. When the student does not know what the project is about, the decision seems easy. However, be sure that you are not mistaking nervousness or shyness for ignorance. When you come across a project that is obviously the work of a student and is original, simply stated, well-conceived, clearly displayed, and well-understood, you have a winner! Some projects will seem far too sophisticated to be the work of a student of the age level you are interviewing. That hunch merits further questioning but not a conclusion. Some students are extremely bright. Others have helpful parents who willingly, and for the best of educational motives, ensure that their child has many resources and contacts available to them. You must distinguish between the student with resources who used them with comprehension and the student who just used them. Although child prodigies do exist, it is unlikely that 10 of them would be placed side by side in the section you are judging. Your task is to make sure that all the real scholars are recognized and rewarded.

Developing Scientific Literacy

We hope that this fair will promote each student's maturation into a systematic problem-solving adult. Some of them may even be a step closer to entering a science or science-related career as a result of this experience. We hope that each develops into a citizen who can critically evaluate research claims as reported in advertisements and the media and into a citizen who values and supports basic scientific research.

We trust that you will find much personal and professional satisfaction in helping to make this happen. Your satisfaction as a judge will be directly proportional to the extent you find fulfillment in assisting youngsters with varying abilities and degrees of sophistication. You have succeeded if students have been encouraged to investigate phenomena more systematically and have been appropriately awarded recognition for their degrees of success. There are no losers at this event, only winners. The losers are those who did not try.

We look forward to meeting you on what will be a most exciting day for everyone involved. Thank you again for agreeing to be a judge in our science competition.

Sincerely,

Julia H. Cothron
Ronald N. Giese
Richard J. Rezba

Related Web Site

http://www.eduzone.com/Tips/
  science/SHOWTIP2.
  HTM (Judges Ballot)

CHAPTER 20

Maximizing Student Success

Objectives

- Distinguish between redesigned and traditional science competitions and cite positive and negative features of each.
- Describe a redesigned science competition based on criterion-referenced measures.
- Describe science competitions available to students.
- Prepare students to participate in redesigned and traditional science competitions.
- Develop a personal philosophy regarding appropriate types of science competitions for your students.

The major function of elementary and secondary schools is to transmit our culture to the young. Science teachers are responsible for conveying our scientific heritage; a heritage that includes vast amounts of content—as well as skills. Included also are the intellectual attitudes and skills involved in applying knowledge and skills to new and unresolved problems. Too often, however, the total focus of science instruction is on the first two of the three aspects of our heritage. Students spend 12 to 16 years learning the history and rules of science, but very few get a chance to actually solve problems in the classroom or to participate in competitive events. To fully benefit from participation in science, students must learn the skills of scientific investigation, regularly practice the skills with classroom investigations, and participate in competitions.

ASSUMPTIONS

An assumption of this book is that students will understand scientific facts, theories, and concepts at a deeper level if laboratory exercises and science competitions develop the research and communication skills used in the advancement of science. Another assumption is that it is important to practice the skills of scientific investigation in practice sessions with regular laboratory exercises and in public forums. Furthermore, it is assumed that in any competition all contestants must play by the same rules and that the judges must be knowledgeable of these rules and must administer them fairly. A final assumption is that alternative models for student participation in competitions should be explored, including criterion-referenced models as well as the typical competitive or norm-referenced models.

RULES OF THE COMPETITION

A first and most basic question that any competition must answer is what is the purpose of the competition and what kinds of projects will be allowed to compete. Should we allow models, demonstrations, and descriptive studies or must they all be experimental in nature? Must the projects be independently conducted or can they be implemented by students working in groups? Should we distinguish between students who have professional mentors and those who work independently? Must students do all of the laboratory work themselves or may they make use of private or state laboratories for certain procedures? Should students from magnet schools compete against students from regular schools?

Once decisions have been reached on the purpose of the competition and on the participants, the competition's rules and judging criteria must be established. Judging cannot be consistent or focused when the rules are defined by nebulous categories. In many science competitions the rules themselves are enough to cause a process-product conflict among judges. Some judges focus on the scientific sophistication and correctness of the project. Other judges focus on the correctness of procedures used to resolve the problem. This issue is one that has dogged competitions for years. It is serious enough to cause many to question the very existence of science competitions as being inherently unfair and inconsistent.

Juliana Texley (1988), a former editor of *The Science Teacher*, cited the problems of (1) students with the resources of laboratories and mentorships competing against students who lack such resources; (2) judges who understand students but lack scientific expertise; and (3) judges who understand science but not students. Such conflicts do not have to be inherent in competition; however, they are inherent when loose rules can be interpreted to stress either process or product over the other. A major step in improving many competitions would be the distribution of a complete set of the competition's rules and criteria to every teacher and judge. A summary of these rules and criteria should also be distributed to students at the start of their projects. Far too often it is assumed that everyone knows the rules and how they are to be interpreted.

JUDGING COMPETITIONS

Retired players are not automatically good referees. Similarly, practicing scientists need to be explicitly reminded of the rules of the competition and the level of the students' sophistication. Otherwise, they will tend to judge from their current positions and the following all too typical remarks are heard. "Not enough statistical analysis," says the statistician of a seventh grader's project. "The model is not wide enough to observe all the real-world effects of the wave on a sea wall," says the marine engineer. "It would have been better to study the interactive effect of those three variables at once," says the senior researcher.

Eliminate the possibility of these scenarios by carefully training judges to administer the rules and criteria you have established for your competition. Suggestions for training judges are provided in Chapter 19, *Preparing to Judge Competitions*. Monitor the judge's performance to insure that rules have been followed. More importantly, remove any judge from future competitions whose interactions with students are negative or inappropriate for the student's age. Lasting negative self-concepts and images of science are formed when judges make inappropriate comments to students. Not only does the competitive event receive a bad reputation, but science may lose the talents of promising individuals.

ALTERNATIVE COMPETITIONS

One option for resolving the process-product dilemma in science competitions lies in the model used to judge figure skating. First, the contestants are judged on compulsory figures that must contain many explicitly stated elements done in a fixed format. Then the competitors are judged on an independently developed routine that allows for much more creativity and uniqueness, but that still has some specific regulations. Competitors are rated after they have finished their compulsory figures and after their unique routine. Based on a combined score of their mastery of the basic skills and their application of these skills to an original routine, trophies are awarded. Similarly, students can be judged with a two-step process.

Step One: Judge on a strictly criterion basis for the scientific research process. Did the student clearly and explicitly conduct a research investigation in which all of the components of experimental design and data analysis were explicitly stated? On the basis of this level of judging of the competition, certificates can be awarded to signify the level of technical problem solving competence that is demonstrated.

Step Two: Judge the sophistication, the intellectual creativity, and uniqueness of the projects. Then based on a combined score of mastery of skills and uniqueness of design, awards can be given. Because ideas evolve over time, their development can be traced in the student's log and further explored during the student's interview. Requirements for the student's log can be altered to include documentation of ideas and variations, as well as raw data.

The Science Project Evaluation Form provided in Chapter 19 can be modified to reflect the two-step process.

Step One: Scientific Research Process

- Background Knowledge: Key scientific concepts, literature review
- Experimental Design: Hypothesis, independent variable, dependent variable, constants, control, repeated measures
- Materials and Methods: All needed materials listed; clear and precise procedures
- Results: Data tables, graphs, summary sentences or paragraphs, support of hypothesis
- Conclusion: Major findings, interpretation, suggestions for further study, application
- Interview: Communicates scientific basis, describes design principles, explains data analysis and recognizes limitations

Step Two: Scientific Product

- Creativity: Unusual question, synthesis of background information, use of materials, and so on

- Sophistication: Unusual research design, knowledge of subject relative to age and school resources, etc.
- Display: Attractive and legible, accurate, consistent with the competition's regulations
- Interview: Communicates source of ideas, evolution of project over time, influence of other individuals, future implications

Similarly, the questions suggested for judges to ask in the interview can be expanded to include questions related to the product.

- What gave you the ideas for the project?
- How did your ideas about the project change over time?
- Who was helpful in providing ideas or helping you refine your ideas?
- How do you plan to apply or expand your ideas?

MAJOR SCIENCE COMPETITIONS

There are many competitive science events. The question is "How does one become aware of the rules and the details of involving students?"

Begin by collecting information about science competitions from your department head or science supervisor. Call a science education professor at the nearest university or the state science supervisor in the Department of Education in your state capital. Ask them for information and the addresses of contact persons for various science competitions.

Browse recent issues of the NSTA publications—*The Science Teacher, Science Scope,* and *Science and Children.* Frequently there are articles and announcements concerning upcoming science competitions. Also check the programs of local, state, and national science teacher association meetings for presentations related to science competitions. Often such presentations are made by teachers and science supervisors who have been successful in motivating students to enter science competitions or by a representative of the competition itself. There is no substitute for being professionally active to receive and share information.

Following is a listing of some of the major student science research related competitions.

American Junior Academy of Science (AJAS) and Individual State Junior Academies of Science (State/National) At the state level students whose research papers are selected to make oral presentations of their research. Winners at the state level are then invited to present papers at the national AJAS competition. For more information contact:

Ms. Gloria Takahashi
Director, AJAS
Southern California Association of Science
900 Exposition Blvd.
Los Angeles, CA 90007

Individual state organizations of AJAS may be contacted by searching the world wide web for the individual state academy of science.

Craftsman/NSTA Young Inventors Awards Program (National) Open to students grades 3–5 and 6–8 who must invent or modify a tool in order to perform a practical function.

Craftsman/NSTA Young Inventors
Award Program
NSTA
1840 Wilson Blvd.
Arlington, VA 22201-3000
Phone   1-888-494-4994
email: younginventors@nsta.org
http://www.nsta.org/program/craftsman.shtml

Duracell Scholarship Competition (National) Open to students grades 6–12 who must design and build working devices powered by Duracell batteries. Students must submit a photograph of their self-contained working device they designed and constructed, in addition to a one to two page description and a wiring diagram of the device. For further information contact:

Duracell/NSTA/Scholarship Competition
1840 Wilson Blvd.
Arlington, VA 22201-3000
Phone   1-888-494-4994
http://www.nsta.org/programs/duracell.shtml

Estes Rocketry Competition Students submit their rocket design(s) for quarterly contests. For more information contact:

Matt Steele
Estes Rocketry Contests
Estes Industries
1295 H Street
Penrose, CO 81240
719-372-6565
http://www.service.com/estes/
teacher~resources.html

Intel International Science and Engineering Fair Open to students grades 9–12 who can compete in 15 different areas of science, mathematics, and engineering in order to win not only scholarship and tuition grants but also scientific equipment or scientific field trips for their entire school. Students' original math, science, and engineering research projects are judged on the basis of both a written and oral research report and exhibit. For more information contact:

Intel International Science and Engineering Fair
Science Service
1719 N Street, NW
Washington, DC 20036
202-785-2255
http://www.sciser.org/weststs.htm

Junior Engineering Technical Society, JETS (Local/Regional/National) Special awards are given at the International Science and Engineering Fair and affiliated regional fairs. JETS also sponsors clubs as well as individual and team related competitions. For more information contact:

JETS Teams
1420 Kings St. Suite 405
Alexandria, VA 22314-2715
Phone   703-548-5387
Fax   703-548-0769
email   jets@nas.edu
http://www.asee.ort/jets/teams/

Junior Science and Humanities Symposium (JSHS) (National/International) Students submit papers, 1,200 words, describing their research. Students selected on the basis of the research paper will be invited to make oral presentations of their work at the symposium. For more information contact:

JSHS Office
The Academy of Applied Science
PO Box 2934
Concord, NH 03302-4520
603-228-4520
http://www.jshs.org/NATIONAL.htm

National Junior Horticultural Association Projects (State/National) Students enter their horticultural project in a state Junior Horticultural Association Contest. Winners advance to national finals. For further information contact:

Jon Hoffman
National Junior Horticultural Association
441 E. Pine Street
Fremont, MI 49412
616-924-5237
http://www.ces.nesw.edu/depts/hort/hic/
enroll.htm

National Science Olympiad (Local/Regional/State/National) School teams compete in a series of science knowledge and skill competitions. For more information contact:

National Office Science Olympiad
5955 Little Pine Lane
Rochester, MI 48306
248-651-4013
248-651-7835
http://www.geocities.com/capecanaveral/
lab/9699

NSTA Standing Committee on Student Programs This website lists 31 competitions.

http://www.nsta.org/pressrel/
studentcompetitions.htm

Odyssey of the Mind (Regional/State/National) Teams of students enter and present solutions to various novel problems. Teams with winning entries are invited to attend a competition in which they are asked to find solutions to other novel problems. For information contact:

Odyssey of the Mind
PO Box 27
Glassboro, NJ 08028
http://www.odyssey.org/

ScienzFair™ Project Competitions Check this web site for nine separate competitions: Balloon Car, Hot Air Balloons, Bottle Rocket, Hot Air Balloons II, Solar Ovens, Bridge Building, Inventions, Catapults, and Spaghetti Bridge.

http://www.members.aol.com/ScienzFairs/
compete.htm

Space Science Student Involvement Project, SSIP (Regional/National) National competition sponsored by NASA and NSTA open to students grades 3–12. Offers a large variety of competitions that focus on math, science, and technical literature. Student may participate in a variety of mediums including art, writing, and science.

Wendell Mohling
NSTA
Space, Science, and Technology Department
1840 Wilson Blvd.
Arlington, VA 22201-3000
703-243-7100
http://www.nsta.org/programs/ssip.htm

IMPROVING COMPETITIVE EVENTS

Events at the local, regional, state, and national level vary in quality from those that exhibit many of the pitfalls previously described to those that are well organized and implemented. Regardless of quality, all events can be improved. Do your part by becoming involved.

- Participate now. No event or competition is perfect. Work for the improvement of the competitions your students enter. Waiting until things become perfect robs your students of the experience.
- Encourage those in charge to ensure that judges, teachers, parents, and students know and follow the rules.
- Recommend that judges be informed, trained, and monitored.
- Urge that the emphasis be shifted from winning to learning the skills of investigating and to experiencing the joys of scientific discovery.

Appendix A: Using Technology

This appendix contains four resources:
I. Steps for graphing data on Texas Instrument's TI-83 graphing calculator.
II. Steps for graphing data on Casio's 9850 G+ graphing calculator.
III. TI-83 calculator program for a one-way Chi-square analysis. (See Chapter 11)
IV. Casio G+ calculator program for one-way Chi-square analysis. (See Chapter 11)

Note: There are often multiple ways to accomplish the same result on a calculator. The following steps are but one way. For other ways, consult the manual that accompanied the calculator.

I. Steps for graphing data on Texas Instrument's TI-83 graphing calculator.

1. Press the **ON** key

2. *Check the MODE* (setting how numeric entries are interpreted and displayed)
   * Press **MODE** key, check that all functions are highlighted to the left.
   * If not, arrow down to highlight and change each line as needed and press **ENTER** for each change.

3. *Clear lists* (removing old data)
   * Press the **STAT** key.
   * Press a **1** on the keyboard (for 1:edit) or simply press **ENTER**.
   * If the lists have old data:
     * Arrow to the top of a list to highlight.
     * Press **CLEAR** and then, **ENTER**.
     * Repeat for each list.

4. *Clear Y =* (removing old equations)
   * Press the 'Y=' key.
   * Move blinking cursor to the 'Y=' to be cleared.
   * Press **CLEAR** to remove old equation.
   * Arrow down to the other 'Y=' and press **CLEAR** to remove.

5. *Set the STAT PLOT* (defining how to plot statistical data)
   * Press **2nd** key and then the 'Y=' key.
   * To define plot 1, press **ENTER**.
   * Use arrows to highlight desired selections, press **ENTER** after each selection.
   * Selections are:
     * On or off
     * Type of graph (scatter, line, bar, box-and-whisker)
     * X list for independent variable
     * Y list for dependent variable
     * Type of mark (squares, crosses, dots)

- To define additional plots, arrow up and across to highlight plots 1, 2, or 3 and press **ENTER**.
- Use arrows to highlight selected plot number and press **ENTER**.
- Highlight desired selections as previously described.
- To turn on or turn off *all* plots, highlight numbers 4 or 5, press **ENTER**.

6. *Set window* (defining the viewing window, i.e., setting the interval scales for the axes)
   - To do this automatically, press **ZOOM** and then **9**.
   - To manually determine the boundaries and other attributes of the viewing window, see 'defining the viewing window' of the TI-83 guidebook.

7. *Graphing the data* (displaying the data)
   - Press **ZOOM 9** and you should see a graph of your data points.
   - Or set window manually and press the **GRAPH** key.

8. *To draw a line-of-best fit, use a mathematics technique called regression*

   If the data looks straight, try a linear regression analysis.
   If the data looks curved, try a quadratic analysis or exponential or power regression.

   - Press the **STAT** key, and arrow over to **CALC**; then arrow down to your choice, e.g., Linear regression [Lin Reg (ax+b)] and press **ENTER**.
   - Your choice will appear on the screen, e.g., LinReg (ax+b).
   - Tell the calculator where the data are, e.g., if the data are in lists 1 and 2, press **L1** (press 2nd key and 1), then press **comma**, then **L2** (press 2nd key and 2). Your screen should look like: LinReg (ax+b) L1,L2
   - To copy the resulting equation to a Y=, follow the steps below:
     - Place a comma after L2 so that the screen looks like: LinReg (ax+b) L1,L2,
     - Press the **VARS** key, and arrow over to **Y-VARS**.
     - Press **ENTER** or **1** to select **FUNCTION**.
     - On the **FUNCTION** menu, select a Y= (e.g., Y1) and press **ENTER**. Your screen should now look like: LinReg (ax+b) L1,L2,Y1 and a blinking cursor. Press **ENTER** and the calculator will calculate the equation and will also paste a copy in the designated Y=. That equation in 'Y=' will allow you to superimpose a line-of-best fit over your data points.
     - Press the **GRAPH** key and a line-of-best fit should appear on your graph.
     - Press the **Trace** key to trace the points of the graph. To trace the line-of-best fit, press the 'up' arrow to shift the cursor to the equation line-of-best fit. Moving the cursor along the line will allow you to predict values not directly measured.

II. **Steps for graphing data on Casio's 9850 G+ graphing calculator.**

1. *Access the MAIN MENU*
   - Press the **MENU** key.

2. *Choose the STAT icon*
   - Use the circle of arrow keys to highlight **STAT** (for statistics), press the blue **EXE** key.
   - Or simply press #2 on the keypad.

3. *Set the window to automatic*
   - To automatically have the calculator select the appropriate window (interval scales) for the graph, press **SHIFT,** then **MENU** to access SETUP.
   - 'Stat Wind' refers to statistics window, which you can set as automatic or manual.
   - Note the row of 'F' keys below the screen. These keys control what appears along the bottom of the screen. Press **F1** to select automatic if necessary.
   - Press the **EXIT** key to return to the lists.

4. *Delete old data from lists*
   - Use the **F6** key to access a screen-bottom menu that includes **DEL-A** [for delete all].
   - To delete old data in List 1, for example, use the arrow keys to highlight anywhere in List 1 and press **F4, DEL-A** [for delete all]; press **F1** [for yes].
   - Arrow over to List 2 and repeat steps as necessary to delete data.

5. *Enter independent variable data in list 1* (e.g., drop height data)
   - Press **EXE** after each entry.

6. Enter dependent variable data in List 2 (e.g., bounce height data); check that you have correctly entered all data.

7. *Begin to graph the data*
   - Press **F6** to access a screen-bottom menu that has **GRPH** [for graph] above **F1**; press **F1.**

8. *Set up StatGraph 1 for a scatter plot*
   - Press **F6** [for SET] to make graph choices.
   - Arrow down to *Graph Type* and use the **F** keys to select **F1, Scat** [for scatter plot].
   - Arrow down to make other choices: XList (1), YList (2), frequency (1), mark type, and color.
   - To simultaneously set up additional graphs, arrow back to the top and highlight StatGraph 1. Use the **F** keys to select GPH2 or GPH3.
   - **EXIT** out when finished setting up the graph by pressing the **EXIT** key.

9. *Graph the data*
   - From this bottom menu, press **F1** to see the graph [GPH1].
   - If you have set-up two or three graphs, press **F4** [for SEL select] and use the **F** keys to turn *on* or *off* Graphs 1, 2, and 3 as desired. Then press **F6** [to DRAW the graphs].

10. *Draw a line-of-best fit on a scatter plot (so you can see the general trend of the data)*
    - From the bottom menu, press **F1** [for x, linear regression] for example.
    - From this screen, press **F5** to copy the equation to an empty 'Y=', press [**EXE**] to store.
    - Then press **F6** to draw a line over the scatter plot. See Step 11 to trace this equation line.

11. *To trace a line-of-best fit (so you can predict other values)*
    - After completing STEP 10, press **MENU** and select Graph.
    - Press **F6** (DRAW) to view the equation line-of-best fit.
    - Press **SHIFT F1** to access the Trace Function.
    - Use the left and right arrows to move the cursor along the line to predict new value for y.

Note: To display other types of graphs, select different graph types in *Step 8.*

III. TI-83 calculator program for one-way (one-independent variable) Chi-square analysis. (See Chapter 11)

ClrList (L$_1$)
ClrHome
Disp "ENTER NUMBER"
Disp "OF CLASSES"
Input N
For(I,1,N,1)
ClrHome
Disp "CLASS"
Disp I
Disp "NUMBER OBSERVED"
Input O
Disp "NUMBER EXPECTED"
Input E
(O-E)^2/E→L$_1$(I)
End
ClrHome
Disp "CHI SQUARE VALUE"
Disp ""
Disp sum(L$_1$)
Disp ""
Disp ""

IV. Casio G+ calculator program for one-way (one-independent variable) Chi-square analysis. (See Chapter 11)

"ENTER NUMBER"↵
"OF CLASSES"?→N↵
N→Dim List 1↵
For 1→I To N Step 1↵
ClrText↵
"PRESS [EXE] TO"↵
"ENTER DATA FOR"↵
"CLASS":I◢
"NUMBER OBSERVED"?→O↵
"NUMBER EXPECTED"?→E↵
((O-E)^2)÷E→A↵
A→List 1[I]↵
Next↵
ClrText↵
"CHI SQUARE VALUE"↵
""↵
(Sum List 1)◢
""↵

··· Appendix B: Practice Problem Answer Key ···

CHAPTER 1

Floor Wax Test

1. Independent Variable: Brand of wax
2. Dependent Variable: The number of scratches
3. Constants: Floor test section size; amount of wax; amount of time test lasted; same hall in same mall.
4. Repeated Trials: 5
5. Control: Floor sections receiving no wax
6. Title: The Effect of Using Different Floor Waxes on the Number of Scratches in Floor Tile.
7. Hypothesis: If the grade of floor wax is varied, then "Tough Stuff" floor wax will result in fewer scratches in the tile than "Steel Seal" floor wax.

CHAPTER 2

Scenario 1

Title: The Effect of the Amount of Water Applied on Plant Height
Hypothesis: If the amount of water applied to plants is increased, then the height of the plants will increase.

IV: Amount of water (ml)				
50 ml	100 ml	150 ml (control)	200 ml	250 ml
10 seeds	10 seeds	10 seeds	10 seeds	10 seeds

DV: Height of plants

C: 500 g soil; Peat's Potting Soil
Distilled water
Watering frequency
Number of seeds in each pot
Time (40 days)
Improvements: Specify a single seed species, variety, brand, age, and so on; all pots must be identical; specify growing conditions: light, temperature, location; make more measurements of the DV: for example, measure the plants every 5 or 10 days, then operationally define the measurement of height (cm).

Scenario 2

Title: The Effect of the Color of Mashed Potatoes on Their Selection by Kindergartners
Hypothesis: If potatoes are colored red, green, yellow, and blue then kindergartners will choose potatoes in the following order: red, yellow, blue, green.

IV: Colors of mashed potatoes			
Red	Green	Yellow	Blue
100 student choices	100 student choices	100 student choices	100 student choices

DV: Students choice of color of mashed potatoes

C: Identical bowls
Mashed potatoes
Scoop size
Improvements: Add a control—a bowl of regular, white potatoes; change the order of the colors each day.

Scenario 3

Title: The Effect of the Height of a Hole on the Side of a Carton on the Distance Liquid Will Squirt
Hypothesis: If the height of a hole on the side of a carton is increased, then the distance a liquid will squirt before hitting a surface will increase.

IV: Height of hole			
5 cm	10 cm	15 cm	20 cm
1 carton	1 carton	1 carton	1 carton

DV: Distance liquid squirted (cm)

C: Identical, quart-sized milk cartons; hole size; height of liquid
Improvements: Conduct more trials; specify the liquid (What is it? What is its temp.? and so on); specify hole size and shape; identify the control.

Scenario 4

Title: The Effect of the Number of Seeds Planted in a Container on Plant Appearance
Hypothesis: If the number of seeds planted in a container increases, then the plants look less healthy.

IV: Number of seeds				
2	4	8	16	32
1 cup	1 cup	1 cup	1 cup	1 cup

DV: Appearance of plants

C: Plastic cups
Amount of soil
Improvements: Use only one kind of seed and specify seed species and variety, and so on; need to identify a control; need to specify size, composition, and color of cup; need to specify type and amount of soil; need to better define plant appearance: for example, height, color, number of leaves, size of leaves, stem diameter; need more trials; need to specify lighting conditions, temperature conditions, and watering procedure.

Scenario 5

Title: The Effect of the Type of Insulation Wrap on the Temperature of Water in a Jar

Hypothesis: If jars of water are wrapped with different types of insulation, then the temperature of the water in the jars will increase by different amounts.

IV: Kind of insulation				
A	B	C	D	E
1 jar	1 jar	1 jar	1 jar	1 jar

DV: Temperature of water in jar

C: Jars all 1/2 full
Jars placed in direct sunlight
Jars fitted with plastic lids
Improvements: Add a control—unwrapped jars; add repeated trials—for example, 5 jars with each type of insulation wrap; specify size, shape, color of the identical jars used; specify the initial temperature of the water used; specify the amount of time the jars are in the direct sunlight; specify the time of year and day that the jars are in the direct sunlight.

CHAPTER 3

1. Selected Sample Answers

A. Meal worms

Question 1.   Materials available: Meal worms, water, food, light, temperature

Question 2.   Action: Meal worms crawl.
Meal worms reproduce.   Meal worms react to stimuli.
Meal worms metamorphose.
Meal worms eat.

Question 3.   Change to affect action:

Meal worms	Light	Meal	Water	Temperature
species	color	composition	amount	variability
size	duration	texture	source	heat source
age	intensity	non-foods	substitutes	
	shade	age	humidity	

Question 4.   Response:
Number of larvae, pupae, adults
Number of meal worms at specific sites
Size of larva
Rate of development
Number of new larva

B. Insect repellent

Question 1.   Materials available: Skin, insects, repellent

Question 2.   Action: Repellents repel insects.

Question 3.   Change to affect action:

Skin	Repellent	Insects
part of body	brand	species
gender of person	amount	stage of development
age of person	frequency	habitat
color		
dryness		
covering		

Question 4.   Response:
The number and kind of insects that land on surfaces
Count number of bites
Distance from sprayed area that insects land or bite

C. Molds

Question 1.   Materials available: Food, environmental conditions

Question 2.   Action: Molds grow.   Molds die.

Question 3.   Change to affect action:

Food	Environmental Conditions
kind	temperature
texture	humidity
acidity	enclosure
exposure	light—color, intensity, source, and duration
age/freshness	
wetness	
location	

Question 4.   Response:
Color
Growth rate
Size
Shape

N. Ice cubes

Question 1.   Materials available: Ice, containers, water, insulating material

Question 2.   Action: Ice cubes melt.   Water freezes.

Question 3.   Change to affect action:

Ice	Containers	Water	Insulating Materials
size	size	composition	composition
shape	shape	additives	texture
composition	composition	other liquids	temperature
color			state
			color
			amount

Question 4.   Response:
Time to change state
Change in volume
Change in shape
Change in mass

O. Paints

Question 1.   Materials available: Paints, surfaces

Question 2.   Action: Paints cover surfaces.   Paints run/drip.   Paints mix.

Question 3.   Change to affect action:

Paints
color
brand
composition
method of application
mix

Surface
composition
texture
size
orientation

Question 4.   Response:
Density of cover
Amount of paint before first drip or run
Amount of paint to cover a surface
Amount of paint to cover a certain area
Ratio to produce a certain shade

2.   Selected Sample Answers

List A
(Possible General Question: How much can the water heap above the rim of a glass?)

Question 1.   Materials available: Glasses, water, coins, soaps, hot plate, salts

Question 2.   Action:

Water heaps.   Water wets objects.
Salts dissolve.   Water flows.
Soaps dissolve.

Question 3.   Change to affect action:

Glasses	*Water*	*Soap or Salt*	*Coins*	*Hot Plate*
volume	type	brand	size	other heat
depth	temperature	type	composition	sources
diameter	substitutes	amount	method of	temperature
shape	source		putting in water	time
composition				
lip shape				
texture of lip				

Question 4.   Response:
The number of coins added to the level glass of water before it spills over
The height of the heap above the rim
The amount that spills over when the surface breaks

List D
(Possible General Question: How can fruit be best packed for shipping?)

Question 1.   Materials available:

Peat moss, wood and metal, blocks, newspaper, sand, water, acid, fruit, salt, flat containers

Question 2.   Action: Fruit bruises.   Fruit spoils/decays.

Question 3.   Change to affect action:

Peat moss/ Soils/Sand	Blocks	Water	Flats	Fruit	Acid
amount	size	amount	size	kind	concentration
compaction	shape	how applied	shape	arrangement	kind of acid
wetness	composition	when applied		number	when applied
depth	arrangement			ripeness	how applied
arrangement					

Question 4.   Response:
Number with bruises
Number of spoiled pieces of fruit
Total number of bruises
Severity of bruising

3.   Selected Sample Answers

Pier Point City

Question 1.   Materials available: Time, lures, fish, body of water

Question 2.   Action: Fish are caught.

Question 3.   Change to affect action:

Time	Lures	Fish	Body of Water
time of day	kind	kind	temperature
time of year	action	age	degree of shade
	shape	size	depth
	color		distance to shore
	company		bottom type
	natural/artificial		vegetation
	size		

Question 4.   Response:
Number of fish caught
Size of fish caught
Kind of fish caught

Fiberville

Question 1.   Materials available: Threads, binding

Question 2.   Action: Twine breaks.   Twine stretches.

Question 3.   Change to affect action:

Threads	Binding
composition	type
thread patterns in twine	application process
number of threads in pattern	amount
thickness of threads	concentration
number of strands in threads	
age	

Question 4.   Response:
Maximum weight supported by twine
Maximum sudden force before breaking

CHAPTER 4

1. **Selected Sample Answers**

A. Measure 50 ml of lukewarm water in a graduated cylinder. Pour the 50 ml of water into a beaker. Place the beaker on a wire gauze pad which is supported on a ring stand or tripod. Measure the initial temperature of the water. Place a candle under the wire pad. Light the candle. Stir the water slowly. Measure and record the temperature of the water every 30 seconds. Repeat the sequence twice more for a total of 3 repeated trials.

2. **(Example from Chapter 3 Problem 2 List A)**

Title: The Effects of Different Solutes on the Extent That Water Will Heap
Hypothesis: If different solutes are added to water, then the extent to which water will heap will decrease.

IV: Different Solutes			
Distilled Water *(Control)*	Liquid Dishwashing Detergent	Liquid Laundry Detergent	Mouthwash
4 glasses	4 glasses	4 glasses	4 glasses

DV: Amount water heaps (the number of coins added to glass of water before it spills)
C: 4 identical glasses
Amount of water
Temperature of water
Amount of solutes
Use of liquid solutes

Sample Procedure

1. Place 4 identical glasses on a flat surface.
2. Fill one of the glasses with 20°C distilled water until you can see the water above the rim of the glass.
3. Place a straight edge of a ruler or a 3 X 5 card level on the rim of the glass.
4. Move the card or the ruler across the rim of the glass.
5. Hold a coin on the edge. Carefully lower it vertically into the glass of water.
6. Add pennies until the water spills over the side of the glass. Record the number of pennies added.
7. Repeat steps 2 through 6 for the other 3 glasses of water.
8. Measure 5 glasses of distilled water into each of 3 large containers.
9. Add 8 tablespoons of liquid dishwashing detergent to one of the large containers of water. Label it. Add 8 tablespoons of liquid laundry detergent to another of the containers of water. Label it. Add 8 tablespoons of mouthwash to the last container of water and label it.
10. For each of the solutions prepared in step 9, repeat steps 1–7.

CHAPTER 5

1. A. Bar graph only
    B. Line Graph
    C. Bar graph only
    D. Line graph

2A. The Effect of Depth on the Number of Fossils Found

Distance below surface (cm)	Number of fossils
80	0
140	2
200	8
260	15
320	32

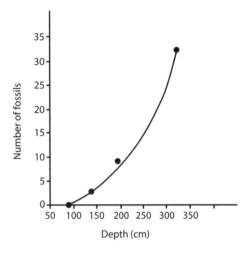

2B. The Value of Food Products Imported in 1988

Imported food	Value ($ billions)
Beef & Veal	1.7
Cheese	0.4
Coffee	2.5
Grapes	0.3
Orange Juice	0.6
Pork	0.9
Shellfish	2.7
Tomatoes	0.2

(Order determined by alphabetization of independent variable. Don't allow students to order dependent variable!)

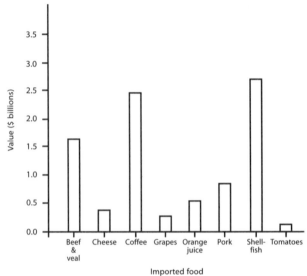

2C. The Effect of Age on the Number of Situps That Males Can Do

Age in years	Number of situps
14.0	93
14.5	95
15.0	90
16.0	98
16.5	99
17.0	100
17.5	101
18.0	102

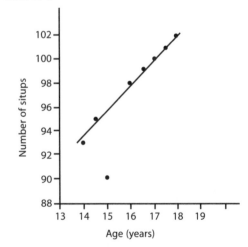

2D. The Effect of Weight on the Number of Situps that Females Can Do

Weight in pounds	Number of situps
100	71
110	69
120	68
130	66
140	64
150	62
160	61
170	59
180	57

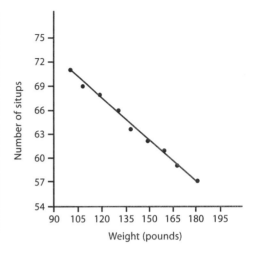

3. a. The number of fossils collected increased greatly with depth.
    b. Of the food products imported into the U.S.A. in 1988, shellfish and coffee had the highest dollar values. Grapes and tomatoes had the least.
    c. As the age of teenage males increases, do does their ability to do situps.
    d. As the weight of female athletes increases, the number of situps they can do decreases.

CHAPTER 8

1. **Heated Soil Scenario**
    a. Quantitative data collected
    b. Interval data
    c. Mean
    d. Range or S.D.
    e. (See table below)

TABLE X.X   The Effect of the Type of Soil on its Temperature Change

Descriptive information	Soil type		
	Sand	**Potting**	**Mixture**
Mean	12	16	11.8
Range	2	4	6.5
Max	13	18	14
Min	11	14	7.5
Number	3	3	3

Peat Moss Scenario
    a. Quantitative
    b. Ratio data
    c. Mean
    d. Range or S.D.
    e. (See table below)

TABLE X.X   The Effect of the Amount of Peat Moss in a Peat Moss/Sand Mixture on Water Retention by the Mixture

Descriptive information	Ratio of sand/peat moss			
	100/0	**80/20**	**60/40**	**40/60**
Mean	73.8	83.6	90	110
Range	10	4	4	12
Max	80	86	90	116
Min	70	82	86	104
Number	5	5	5	5

2. **Deer Mice**
   a. Quantitative data
   b. 1.5 g
   c. 1.5 g – 3.0 g or 1.5 g
   d.
   $$\text{mean X} = \frac{\Sigma x_i}{n} = \frac{15.5}{7} = 2.2 \text{ g}$$

   mode = 2.5 g

   median = 2.5 g

3. A. Nominal data
   B. The Effect of Various Varieties of Apples on Their Ripeness

Descriptive information	Ripeness of apples		
	Variety 1	**Variety 2**	**Variety 3**
Mode	Pink	Yellow-green	Red
Frequency distribution			
Red	60	60	185
Pink	80	50	45
Yellow-green	60	120	30
Dark green	70	40	10
Number	270	270	270

C.

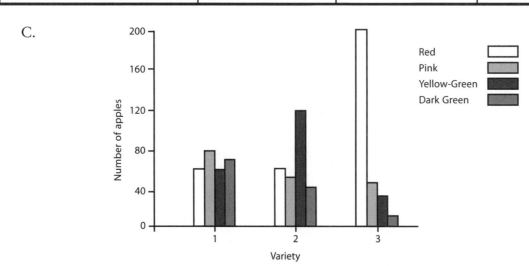

CHAPTER 9

1. Heated Soil Scenario

(Assumed hypothesis: If samples of potting soil, sand and the soil/sand mix are placed under a heat lamp, then the temperature change will be greatest for potting soil, followed by the soil/sand mix and sand.)

The effect of the composition of a soil sample on the temperature change that the sample undergoes when it is exposed to a heat lamp is summarized in Table X.X. The mean temperature change (11.8°C) of the mixture of sand and potting soil was less than the mean temperature change of both the sand (12°C) and the potting soil (16°C). The greatest variation in temperature changes occurred in the mixture with a range = 6.5°C, whereas sand had a range of only 2°C and potting soil 4°C. The hypothesis that the temperature change will be greatest for potting soil, followed by the mix and sand was not supported by the data.

Peat Moss Scenario

(Assumed Hypothesis: If the amount of peat moss mixed with sand is increased, then the mixtures' ability to retain water will increase.)

Summarized data for this study of the effect of various amounts of peat moss in a peat moss/sand mixture on the mixtures' ability to retain water is presented in Table 1.2. As the percentage of peat moss in a peat moss/sand mixture increases from 0 to 60%, the mean amount of retained water also increases. The samples with the lowest (0%) and the highest (60%) percentage of peat moss varied the most, 10 and 12 milliliters respectively. The samples with 20% and 40% peat moss varied least, 4 ml each. The hypothesis that increasing the amount of peat moss in a sand/peat moss mix would increase the mixture's ability to retain water was supported at every level tested.

3. a. Qualitative data
   b. Nominal data
   c. Mode
   d. Frequency distribution
   e. (See table below)

TABLE X.X   The Effect of Darkness of the Shade of Blue Paint on the Sales of a Sports Car to Teenagers

Descriptive information	Blue color of paint				
	VLB	LB	B	DB	DNB
Mode	R	R	B	R	B
Frequency distribution					
Bought	2	3	5	3	7
Rejected	7	6	4	6	2
Number	9	9	9	9	9

Table X.X summarizes the data collected in this study of the effect of the darkness of a shade of blue on the appeal of a sports car to teenagers. In general, as the darkness of the shade of blue increased, the appeal of the car also increased. The exception to this trend was that cars with a dark blue paint appealed only as well as light blue, and less well than blue. Very light blue and dark navy blue had the lowest variation in their appeal to buyers. Cars painted with the intermediate shades of light blue, blue and dark blue showed greater variation in their appeal to buyers. The hypothesis that the darker its shade of blue, then the greater would be the car's appeal to buyers was supported by the data for all the shades tested with the exception of the medium dark blue shade.

CHAPTER 10

2. A

Server 3   J-Shaped

```
2 | 9  9
3 | 0  0  0
3 | 1  1  1  1
3 | 2  2  2  2  2  2  2  2
3 | 3  3  3  3  3  3  3  3  3  3  3  3  3
```

2. B

Server 1   Normal

```
3 | 1  1  1
3 | 2  2  2  2
3 | 3  3  3  3
3 | 4  4  4  4  4  4
3 | 5  5  5  5
3 | 6  6  6  6
3 | 7  7  7
3 | 8  8
```

2. C

Server 4   U-Shape bi-modal

```
2 | 8  8
2 | 9  9  9
3 | 0  0  0  0  0  0
3 | 1  1  1
3 | 2  2
3 | 3  3  3
3 | 4  4  4  4  4  4
3 | 5  5  5
3 | 6  6
```

2. D

Server 2   Rectangular

```
2 | 8  8  8  8  8
2 | 9  9  9  9  9
3 | 0  0  0  0  0
3 | 1  1  1  1  1
3 | 2  2  2  2  2
3 | 3  3  3  3  3
3 | 4  4  4  4  4
```

2. E

	Server 1	Server 2	Server 3	Server 4
range	7	6	4	8
mean	34	31	31.9	32
median	34	31	32	32
Q_1	33	29	31	30
Q_3	36	33	33	34
SD	2.07	2.03	1.27	2.44

2. F   Server 4—The data set has the greatest range and standard deviation.

2. G   Server 3—The data set has the smallest range and standard deviation.

CHAPTER 11

Set 1

1. See definitions in text

2. Sample: The Daphnia taken from grandfather's pond for study
   Sampled population: Daphnia in grandfather's pond
   Target population: All fresh water Daphnia

3. Sample: Apples tested that were randomly selected from the source of apples
   Sampled population: All the apples in the source of Ben's sample
   Target population: All apples

4. Tony's because his findings were based on a larger sample

Set 2

1. a–c. See definitions in text

2. a. 5/100   95/100
   b. 3/100   97/100
   c. 15/100   85/100
   d. 40/100   60/100
   e. 1/1000   999/1000

3. a. Increasing the amount of Chemical X added to water will have no effect on the mean temperature of the solution.
   b. The mean height of loblolly pines near a pollution source of auto exhaust will be the same as the mean height of loblolly pines farther from the pollution source.
   c. The mean blood pressure of stressed people who exercise and those who don't will be the same.
   d. Increasing the slope of a stream ($10°$ to $30°$) will produce no significant differences in the stream's sediment load.

4. a. 2.365   b. 2.878   c. 3.646   d. 2.000

5. a. Reject null/support research
   b. Reject null/support research
   c. Accept null, non-support of research

6. a. The 0.05 level is adequate. The sample size is large, compared to biological studies being investigated. Human health is not an issue.
   b. The 0.001 level is recommended because human health is involved and the sample is relatively small.
   c. The 0.01 level is recommended because the sample size is very small for biological phenomena. Generally, the sample size should be 30 or greater for a statistical test.

Set 3

1. Calculated $t = 0.825$
   df $= 58$, α at 0.05, $t = 2.048$, hypothesis not supported

2. Calculated $t = 6.04$
   df $= 148$, α at $= 0.001$, $t > 3.373$ required
   $\therefore$ differences are significant, reject null hypothesis, support research hypothesis.

3. a. Testosterone and control at df $= 38$, α at 0.001, $t \cong 3.551$, calculated $t = 4.83$ null hypothesis not rejected, research hypothesis supported
   b. Estrogen and control at df $= 38$, α at 0.05, $t = 2.021$, calculated $t = 0.112$ null hypothesis not rejected, research hypothesis not supported

4.

Server 1 & Server 2	Server 1 & Server 3	Server 1 & Server 4
$t = 6.415$	$t = 5.34$	$t = 3.887$
p $= 2.063 \times 10^{-8}$	p $= 1.590 \times 10^{-6}$	p $= 2.635 \times 10^{-4}$
df $= 63$	df $= 58$	df $= 58$
$\bar{X}_1 = 34.27$	$\bar{X}_1 = 34.27$	$\bar{X}_1 = 34.267$
$\bar{X}_2 = 31$	$\bar{X}_2 = 31.9$	$\bar{X}_2 = 32$
S $X_1 = 2.066$	S $X_1 = 2.067$	S $X_1 = 2.067$
S $X_2 = 2.029$	S $X_2 = 1.269$	S $X_2 = 2.435$
S$X_p = 2.047$	S$X_p = 1.715$	S$X_p = 2.259$

Server 2 & Server 3	Server 2 & Server 4	Server 3 & Server 4
$t = -2.101$	$t = -1.806$	$t = .199$
p $= .0396$	p $= .0757$	p $= .843$
df $= 63$	df $= 63$	df $= 58$
$\bar{X}_1 = 31$	$\bar{X}_1 = 31$	$\bar{X}_1 = 31.9$
$\bar{X}_2 = 31.9$	$\bar{X}_2 = 32$	$\bar{X}_2 = 32$
S $X_1 = 2.029$	S $X_1 = 2.029$	S $X_1 = 1.269$
S $X_2 = 1.269$	S $X_2 = 2.435$	S $X_2 = 2.435$
S$X_p = 1.721$	S$X_p = 2.225$	S$X_p = 1.942$

Set 4

1. df $= 6$ α at 0.05 $= 12.592$ for 0–10 m $x^2 = 27.86$ $p < 0.001$ significant
   α at 0.01 $= 16.812$ 11–20 $x^2 = 17.08$ $p < 0.01$ significant
   α at 0.001 $= 18.548$ 20 $x^2 = 17.34$ $p < 0.01$ significant
   research hypothesis was supported

CHAPTER 12

Part I

1. **Title: The Effects of Ultra-Violet Light, Temperature and pH on the Durability of Paint**

IV: UV light	IV: Temp. (°C)	IV: pH of Rain		
		2	4	6
5	10	5	5	5
	20	5	5	5
	30	5	5	5
10	10	5	5	5
	20	5	5	5
	30	5	5	5
15	10	5	5	5
	20	5	5	5
	30	5	5	5

DV: Fading

C: Woodbase
Brand X yellow

2. **Title: The Effect of Color and Time on the Number of Crickets Attracted to a Container**

IV: Dish color	IV: Time (min.)			
	40	60	90	120
Red	30	30	30	30
Yellow	30	30	30	30
Blue	30	30	30	30

DV: Number of crickets at each dish
Number of g mustard seeds consumed in 120 minutes

C: Sunlight condition
Amount of mustard seeds/dish

3. **Title: The Effect of Acidity and Time on Kudzu Stem, Root and Leaf Growth**

IV: pH	IV: Time (weeks)					
	1	2	3	4	5	6*
4	50	50	50	50	50	50
5	50	50	50	50	50	50
6	50	50	50	50	50	50
7 (control)**	50	50	50	50	50	50
8	50	50	50	50	50	50
9	50	50	50	50	50	50

DV: Stem width growth
Leaf width growth
*Root width (measured at end of 6th week)

C: Time
—30 days to start
—washing roots
—1 liter of solution/3 days
—light
—temperature

** pH 7 was picked as control because that is neutral. Either a pH of 5 or 6 could have been chosen instead with the rationale that they are the average pH of rain water.

4. **Title: The Effect of Reward Systems and Time on Cat Maze Walking**

IV: Reward schedule	**IV:** Time (weeks)											
	1	2	3	4	5	6	7	8	9	10	11	12
Regular X	X	X	5	5	5	X	X	X	X	X	X	
Random X	X	X	X	X	X	X	X	X	5	5	5	

DV: Time to walk a maze

C: 3-week learning time, brand of treat, interval from last feeding
3-week rest, cats
maze

(The time sequence for the entire experiment is shown. "X" indicates normal conditions prior to the experiment, no random reinforcement (weeks 4–6), rest (weeks 7–9), and no regular reinforcement (weeks 10–12).

5. **Title: The Effect of the Brand of Water Softener and the Hardness of Water on the Quality of Processed Water**

IV: Brand of softener	**IV:** Hardness of water (mg/l)		
	50	100	150
X	10	10	10
Y	10	10	10
Z	10	10	10

DV: Degree of hardness reduction
Degree of hardness of processed water
Amount of sodium in processed water

C: Water solutions

6. This example involves the correlations between chest girth measurements and weight. First, the researcher examined the correlations across all subpopulations; that is, all deer were grouped together regardless of sex (gender), age, or origin relative to the Blue Ridge Mountains. The experimental design is shown below:

Title: The Correlation of Weight and Chest Girth for Deer

Deer	Chest girth	Weight
1		
2		
3		
.		
.		
.		
etc.		

A scatterplot would be made for the two variables. If a correlation exists, then it would be appropriate to determine if the correlation holds for all subpopulations. The procedure would be repeated using *only* data from a) males, b) females, c) young deer, d) old deer, e) east of Blue Ridge, f) west of Blue Ridge. The purpose of the subpopulation analysis is to determine if any subgroup does not follow the general trend. If so, then chest girth might not be a substitute for field weight measures.

7. **Title: The Effect of Thermal, Phosphate, and Acidic Pollution on the Respiration of Fish**

Acidity pH	10°C		20°C		30°C		40°C	
	0 ppm	3 ppm	0 ppm	3 ppm	0 ppm	3 ppm	0 ppm	3 ppm
6.5	5	5	5	5	5	5	5	5
5.5	5	5	5	5	5	5	5	5
4.5	5	5	5	5	5	5	5	5

Main effects
- Effect of phosphate concentration on respiratory rate
- Effect of pH on respiratory rate
- Effect of temperature on respiratory rate

Double interactions
- Combined effect of phosphate concentration and pH on respiratory rate
- Combined effect of phosphate concentration and temperature on respiratory rate
- Combined effect of temperature and pH on respiratory rate

Triple interactions
- Effect of phosphate, temperature, and pH on respiratory rate

See example on cleaners, p. **141**, three independent variables for data analysis techniques.

Part II

1. **Title: The Effect of Concentration and Time of Application on the Efficiency of Herbicide**

Concentration	Time					
	8 A.M.	**10 A.M.**	**12 P.M.**	**2 P.M.**	**4 P.M.**	
40%	X = 19.2 SD = .45 N = 5	X = 17.2 SD = 1.30 N = 5	X = 9.6 SD = 1.14 N = 5	X = 9.8 SD = .84 N = 5	X = 15.4 SD = 1.14 N = 5	X = 14.4 SD = 4.24 N = 25
20%	X = 17 SD = 1.58 N = 5	X = 12.4 SD = 1.14 N = 5	X = 4.6 SD = .89 N = 5	X = 5.8 SD = .84 N = 5	X = 9 SD = 1.58 N = 5	X = 9.8 SD = 4.75 N = 25
10%	X = 15.6 SD = 1.52 N = 5	X = 7.6 SD = 1.14 N = 5	X = 2.8 SD = .84 N = 5	X = 6.2 SD = .84 N = 5	X = 7.8 SD = 1.30 N = 5	X = 8.0 SD = 4.42 N = 25
	X = 17.5 SD = 2.17 N = 15	X = 12.4 SD = 4.21 N = 15	X = 5.7 SD = 3.11 N = 15	X = 7.2 SD = 2.01 N = 15	X = 10.7 SD = 3.67 N = 15	

The data indicate that for all concentrations, the herbicide was most effective when sprayed in the early morning. It is least effective at noon, and its effectiveness increases in the afternoon, but more slowly than its effectiveness decreased in the morning. For any given time of day, decreasing the strength decreased its effectiveness except when 10% herbicide was used at midafternoon (2 P.M.)

CHAPTER 15 TEST

1. See chapter 2 for definitions

2. Used tea bags under rose plants = IV
   Growth of rose plants = DV

3. Kind of dry cell = Purposely changed variable (IV or MV)
   Distance traveled = Responding variable (DV or RV)

4. 1. g;   2. d;   3. a;   4. c;   5. b;   6. e;   7. f

5. A. independent variable   B. levels of IV   C. number of repeated trials
   D. dependent variable   E. constants

6. IV: Brand of gasoline
   Hypothesis: Sam predicted that Ace gasoline was best
   Repeated Trials: 3 round trips
   Constants: Same make, model, horsepower of autos
                 same route, traffic, speed.
   DV: Miles/gallon
   Levels of IV: Speedy, Ace, Roll-on

7A.

Title: The Effect of Changing the Concentration of Salt Water on the Length of Time It Takes the Water to Cool

Hypothesis: If the concentration of salt in water is increased, then the length of time it will take the water to cool will increase.

IV: Salt concentration in water			
0%	10%	20%	30%
1 Trial	1 Trial	1 Trial	1 Trial

DV: Amount of time to cool to 3°C.

C: Identical plastic glasses
Amount of solution (225 ml)
Same freezer

Ways to improve: Specify 0% solution as the control; more time and temperature measurements as the solutions cool; more trials; specify the brand and type of salt; specify distilled water; specify same initial temperature of solutions.

7B.

Title: The Effect of the Color of a Flower on Its Attractiveness to Bees

Hypothesis: If different colored flowers are present, then bees will be attracted to the flowers in the order of red, pink, yellow and blue.

IV: Color of flowers			
Red	Blue	Yellow	Pink
6 Trials	6 Trials	6 Trials	6 Trials

DV: Number of bees

C: Temperature
Observations for 30 minutes
Location in back yard

Ways to improve: Add a control (white flowers) or specify one color as the control; use only one species of flower; specify weather conditions.

7C.

Title: The Effect of Dropping a Magnet on Its Strength

Hypothesis: If the number of times a magnet is dropped increases, then its ability to attract iron filings decreases.

IV: The number of times magnet dropped			
0 Drops	1 Drop	2 Drops	3 Drops
1 Trial	1 Trial	1 Trial	1 Trial

DV: Amount of iron filings lifted

C: Iron filings
Height of drops
Same cement floor
North end of magnets tested
Same kind of magnets

Ways to improve: Needs to test each magnet with different iron filings of the same size; more trials; test magnets before drops to be sure they are the same strength, identify 0 drops as the control.

8. **Selected Example:** Shadow
Question 1.   Materials available: Light source, object, surface
Question 2.   Action:
Shadows exist
Shadows change.
Question 3.   Change to affect action:

Light Source	*Object*	*Surface*
size	size	texture
distance	shape	color
focus	distance	distance
intensity	sharpness of object	
color	movement	
shape	transparency level	
movement		

Question 4.   Response:
Size of shadow
Shape of shadow
Sharpness of shadow

Selected Example: Birdseed
Question 1.   Materials available: Birdseed, feeder, "scarecrow"
Question 2.   Action:
Birdseeds attract birds.
Birdseeds are eaten.
Birdseeds grow.
Question 3.   Change to affect action:

Birdseed	*Feeders*	*Scarecrow*
brand	size	distance
kinds of seeds	location	color
ratio of kinds	color	size
# of sources	height	movement
color	shape	shape
non-seed foods	platform	
alteration of seeds	support	
	nearness to cover	

Question 4.   Response:
Number of birds
Kinds of birds
Bird behavior
Ratio of kinds of seeds eaten

9. **Selected Example:** Shadow

Title: The Effect of the Angle of the Light on the Sharpness of an Object's Shadow.

Hypothesis: If the angle of the light relative to the object increases, then sharpness of the shadow decreases.

IV: Height of the light source			
0° (Control)	30°	60°	90°
5 Trials	5 Trials	5 Trials	5 Trials

DV: Sharpness of shadow's edge

C: Distance light from object
Intensity of light
Same object
Object and the shadow surface stay in the same position
Same light source

Selected Example: Birdseed

Title: The Effects of Changing the Ratio of the Kinds of Seeds in a Feeder on the Variety of Bird Species that Eat at the Feeder.

Hypothesis: If the ratio of corn to sunflower seeds in a mix is increased, then the variety of birds that eat at a feeder will increase.

IV: Ratio of Corn to Sunflower Seeds				
Regular mix (control) 50/50	60/40	70/30	80/20	90/10
8 trials	8 trials	8 trials	8 trials	8 trials

DV: The number of species at the feeder

C: Time of day observed
Length of observation period
Same size, shape, color, height, and location of feeder

10. & 11.
Answers will vary greatly. Check for the following items: Does the procedure include the following items?

1. States all the materials needed
2. Lists all the necessary steps in sequential order
3. Requires repeated trials
4. Requires a control
5. Includes testing all the levels of the independent variables

(See answers for the practice problems in Chapter 4)

12. a. Line
    b. Bar
    c. Line
    d. Bar
    e. Line

13. **Title: The Effect of Newspaper Promotions on New Subscriptions**

Newspaper	New subscriptions (1989)
Bennington Journal (BJ)	8,500
Florida Gazette (FG)	5,000
New Herald (NH)	42,000
Weekly Press (WP)	40,000
World Yesterday (WY)	10,000

(IV ordered by alphabetization.)

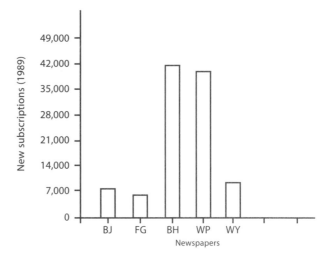

14. **Title: The Effect of Time on the Number of Steps Taken in Physical Therapy**

Time (minutes)		Total steps
9:10 start	0	0
9:15	5	4
9:30	20	9
9:40	30	12
10:05	55	15
10:15	65	20
10:35	85	26
10:40	90	31
11:00	110	36

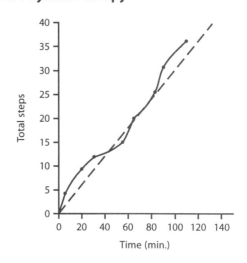

(The solid line is the curve which represents the actual number of steps taken in a given amount of therapy time. Graphs of predator/prey and the changes in the stock market are often graphed this way. The straight line is a line of best fit, and it represents general trend of the data ignoring starts and stops, spurts and rests.)

15. A. The wind speed increases toward the middle of the storm and then dies down again as time passes.
    B. The longer the runner jogs, the slower his pace.
    C. As time went on, the runner ran further.

16. __v__ range   __c__ mean   __v__ frequency distribution
    __n__ sum   __n__ number   __c__ median
    __c__ mode   __v__ variation   __v__ standard deviation

17. Measures of Central Tendency

    __mode__

    __median__

    __mean__

    __mean__

Measures of Variation

    __frequency distribution__

    __frequency distribution__

    __SD/range__

    __SD/range__

18. Height (cm) is quantitative, ratio data because the metric scale for length has an absolute zero and equal intervals.

    Color of leaves is qualitative, nominal data because yellow, green and brown are named categories with no intervals.

19. Grams, as metric units of mass, are quantitative, ratio data because there is both an absolute zero (no mass) and a gram unit has a fixed or definite size.

20. a. Nominal data: days of week, color
    b. Ordinal data: class rank of students, Moh's hardness scale
    c. Interval data: SAT test scores, degrees Celsius
    d. Ratio data: Kelvin Scale of temperature, distance, weight

21A. a. Quantitative
    b. Ratio data
    c. Mean
    d. Standard deviation or range
    e. See table below
    f. See table below

Title: The Effect of the Angle of an Incline Plane on the Speed of a Toy Car

Descriptive information	Ramp angle				
	0°	10°	20°	30°	40°
Mean	0	21.3	12	4	2.7
Range	0	6	4	0	1
Maximum	0	24	14	4	3
Minimum	0	18	10	4	2
Number	3	3	3	3	3

21B. a. Qualitative
    b. Nominal
    c. Mode
    d. Frequency distribution
    e. See table below
    f. See table below

Title: The Effect of Color on the Sales of Roses

Descriptive information	Color of roses sold					
	June 1st	June 8th	June 15th	June 22nd	June 29th	Total sales
Mode	P	P	P	P & W	R	
Frequency distribution						
R	25	15	29	21	30	120
Y	10	20	12	18	10	70
P	30	28	32	24	26	140
W	25	25	16	24	25	115
L	10	12	11	13	9	55
Number	100	100	100	100	100	

21C. a.   Quantitative
    b.   Ratio
    c.   Mean
    d.   Range
    e.   See table below
    f.   See table below

Title: The Effect of Intensity of Lighting on the Number of Brine Shrimp in an Area

Descriptive information	Amount of lighting			
	Area 4	Area 3	Area 2	Area 1
Mean	96.6	74.8	18.4	11
Range	118	26	17	5
Maximum	136	87	28	13
Minimum	18	61	11	8
Number	5	5	5	5

22. Median   24
    Mode   27
    Frequency distribution
    18   1
    19   1
    20   1
    21   21
    22   1
    23   1
    24   2
    25   1
    26   1
    27   3
    28   1
    29   1

23. Mean   24
    Range   11
    Variance = 11.7
    SD =   3.4

24. c.

25. a.

26. b.

27. c.

28. b.

29. d.

30. b.

31.

∠ **0°**

33	2					
34	1	7				
35	3	4	7	9		
36	0	1	2	2	4	8
37	2	3	4	5		
38	2	7				
39	0					

∠ **10°**

34					
35	1	1	2	4	9
36	1	4	5	8	8
37	1	1	3	4	8
38	1	2	4	5	6
39					

∠ **30°**

34	2								
40	7								
41	9								
42	4	4	6						
43	0	1	4	2	3	6			
44	1	1	1	3	5	8	8	9	9

∠ **50°**

38	0	4	5	6	8	9
39	2	2	9			
40	1	4				
41	0	3	6	8		
42	2	3	4	4	6	

32. c

33. d

34. a

35. b

36. 35 ml

37. 37 ml

38. 31 ml

39. a

40. 2.154

41. 6.993

42. Yes

43. $p = 2.215 \times 10^{-9}$

44. a. Sample: 10 students with head lice studied from one school district
    b. Sample population: 100 students with head lice from among whom the 10 were chosen for
                               study
    c. Target population: All students with head lice

45. __4__   Set A
    __5__   Set B
    __2__   Set C
    __1__   No difference between control and test scores
    __3__   Test scores of test group higher than control

46. a. Research hypothesis: If the price of bananas is lower, then they will sell faster.
        Three different comparisons can be made to test this hypothesis:

        • More bananas will be sold at a price of 20¢ than at 40¢
        • More bananas will be sold at a price of 30¢ than at 40¢
        • More bananas will be sold at a price of 20¢ than at 30¢

    b. *t* test
    c. Null hypothesis: The amount of bananas sold is not affected by the price.
        Three different hypotheses can be stated and tested:

        • The number of bananas sold at 20¢ equals the number sold at 40¢
        • The number of bananas sold at 30¢ equals the number sold at 40¢
        • The number of bananas sold at 20¢ equals the number sold at 30¢

    d. See table below
    e. See table below
    f. See table below
    g. Null hypothesis accepted for prices 20¢ and 30¢
        Null hypothesis rejected for 30¢ and 40¢ and for 20¢ and 40¢
    h. Reject the research hypothesis that raising the price from 20¢ to 30¢ will make a difference
        Accept the research hypothesis that raising the price from 20¢ to 40¢ or from 30¢ to 40¢ makes
        a difference

Descriptive information	Price		
	20¢	**30¢**	**40¢**
Mean	22	27	9
Variance	32.5	70	17.5
Standard deviation	±5.7	±8.4	±4.2
1SD	16.3 to 27.7	18.6 to 35.4	4.8 to 13.2
2SD	10.6 to 33.4	10.2 to 43.8	0.6 to 17.4
3SD	4.9 to 39.1	1.8 to 52.2	−3.6 to 21.6
Number	5	5	5

Results of *t* test

     20¢ vs 30¢ $t = 1.1$ no significant difference

     20¢ vs 40¢ $t = 4.3$ $p > 0.01$

     30¢ vs 40¢ $t = 4.1$ $p > 0.01$

At $df = 8$ α of 0.05, $t = 2.306$, and α of 0.01 $t = 3.3555$

47. a.   If the degree of ripeness of bananas increases, then the sale of bananas will decrease
    b. χ^2
    c.   The degree of ripeness makes no difference in the sale of bananas
    d.   Calculated value $\chi^2 = 66.4$
    e.   Yes
    f.   2
    g.   Rejected
    h.   Accepted
    i.   See table below

Title: The Effect of the Degree of Ripeness of Bananas on Their Sales

Descriptive information	Observed distribution	Expected distribution	Calculated χ^2
Mode	yellow		
Frequency distribution			
Green	55	66	1.8
Yellow	110	66	29.3
Black	33	66	16.5
Number	198	198	
Results of χ^2 test $\chi^2 = 47.6$ at df 2 χ^2 of 47.6 > 10.597, p < .001			

48 a.

Title: The Effect of Dilution of Liquid Tylex on the Dissolving of Bathtub Rings
Hypothesis: If Liquid Tylex is diluted, then it will take longer to dissolve a ring around the bathtub and the cleaning water will be cloudier.

IV: Dilutions of Liquid Tylex			
Pure Tylex (Control)	200 ml Tylex 50 ml water	150 ml Tylex 100 ml water	100 ml Tylex 150 ml water
3 trials	3 trials	3 trials	3 trials

DV: Time for bathroom scum ring to disintegrate.
  Cloudiness of cleaning water

C: Tiles, bath water, Liquid Tylex
  Amount of tile in water
  Time in bath water
  Number of times in bath water

  b.   If the concentration of Liquid Tylex is decreased, then it will take longer for the tub scum to disintegrate.
  c.   If the concentration of Liquid Tylex is decreased, then the used cleaning solution will be cloudier.

49. a. mean
    b. standard deviation

50. a. mode
    b. frequency distribution

51. b

52. b

53. c

54. a

55. b

········ Appendix C: Experimenting Safely ········

Wearing your seat belt in a car and using protective pads and a helmet when skateboarding make good sense. Similar safety precautions are also important when conducting a science project. **Before you begin your experiment, be sure your teacher has reviewed your procedures for safety.** If you are conducting your experiment at home, you should also discuss your safety precautions with your parents as well.

Safety concerns for different kinds of projects are described in separate sections of this appendix. These sections are: A) chemicals, B) mold and microorganisms, C) electricity, D) radiation, and E) animals and humans. Read the sections that are related to your project. **The safety guidelines here are only a sample. Be sure you understand and follow all the safety procedures needed for your own project.**

A. CHEMICALS

Cleaners, fertilizers, and other chemicals serve many useful purposes, but all of them can be dangerous if improperly used. Never mix chemicals, not even household cleaners, without help from an adult. In addition, you should:

- **Always wear protective glasses.** Gloves and an apron are also good ideas.
- **Wash your hands after handling any chemical.**
- **Know the potential dangers of the chemical you are using.** Some chemicals can irritate your skin, while others are poisonous.
  Do not breathe in vapors from chemicals.
  Be sure the area in which you are working is well-ventilated.
- **Know how and where to store chemicals safely.** A special kind of container might be needed, or maybe the chemical should be stored in a glass instead of a plastic container.
- **Know what to do in case of an accident.**
- **Know the procedures for safely disposing of your chemicals.**

Your science teacher can help you find answers to safety questions in laboratory manuals or chemical catalogs, such as the *Flinn Chemical Catalog and Reference Manual.* Most schools also have information sheets on the chemicals used in science classes. These are called Material Safety Data Sheets, or MSDS.

B. MOLD AND MICROORGANISMS

You have probably seen mold growing on bread and other foods because molds are all around us. Microorganisms are also everywhere. Most common molds and microorganisms are harmless, but some are harmful. Before beginning any project with molds, ask your parents if you or anyone in your family is allergic to molds. Follow these safety precautions:

- Keep the mold and microorganism containers covered.
- Do not touch the molds or microorganisms.
- Wash your hands frequently.
- Never smell molds and microorganisms by inhaling close to the containers.
- Do not reuse containers.
- Dispose of your organisms and containers properly.
- Avoid growing molds on soil; some can make you sick.

When growing molds and microorganisms, you will often grow "uninvited" molds, bacteria, fungi, and yeasts. Most of these are also harmless, but some are not. Play it safe. Properly dispose of these uninvited guests.

Similar care should be taken when studying other microorganisms such as bacteria, protozoa, and algae. Learn as much as you can about these organisms before beginning any experiment. Bacteria, for example, are often grown in special containers called petri dishes. Harmful bacteria as well as safe bacteria may grow in these containers. Follow the same safety procedures as those given for working with molds.

C. ELECTRICITY

Experiments that use electricity should always be checked by an adult who knows how to safely work with electricity. Take the proper precautions to prevent an accident. When designing and conducting your experiment, you should:

- Use as little voltage as possible.
- Avoid using current from household outlets; use batteries instead.
- Watch for leaky batteries. The chemicals inside can be harmful.
- Make sure electrical appliances and tools are insulated and grounded.
- Never work alone.

D. RADIATION

Experiments using microwave ovens, lasers, radon, and some types of smoke detectors all involve radiation—energy or streams of particles given off by atoms. Radiation can be very dangerous. Even in small amounts it can be harmful to living tissue.

Before beginning any experiment involving radiation, get help from someone who knows about the kind of radiation you would like to use in your experiment. Keep these safety precautions in mind:

- Never work alone.
- Dispose of materials that give off radiation as required.
- Know the law. Certain state and federal laws may apply.

E. ANIMALS AND HUMANS

If you plan to experiment with animals with backbones (vertebrates), you must follow very special rules. Vertebrates include fish, amphibians, reptiles, and birds as well as mammals. If you want to use vertebrates or their eggs, discuss your ideas with your teacher first. A qualified adult supervisor who is trained to take care of vertebrates, like a scientist or a teacher, must agree to begin supervising your project before you even obtain the first organism. A form must be signed by the person agreeing to supervise a project involving vertebrates.

Most schools and competitions prefer that students use animals without backbones (*invertebrates*) in animal experiments. Insects and worms are examples of invertebrates. If you do a project with animals, you must provide proper care for all the animals. Proper care includes:

- **a comfortable living place;**
- **procedures that do not injure the organism;**
- **enough food, water, warmth and rest;**
- **gentle handling;**
- **humane disposal or a proper home for organisms when your experiment is finished.**

If you are conducting an experiment that may be entered in a competitive event, such as a science fair, be sure you read and follow their rules on the use of animals in experiments. See http://www.sciserv.org/isef/form2000.pdf for the most up to date Version of the Intel/Science Service International Science and Engineering Fair rules, or contact the following organization for the cost of a copy.

> Science Service Incorporated
> 1719 N St., N.W.
> Washington, DC 20036
> (202) 785-2255

Special rules must also be followed in experiments using humans. Nothing may be done to humans that is likely to cause them harm. Participation should be voluntary. Some experiments, like those that just involve observing people, may not need special signed forms and procedures. Talk with your teacher about experiments involving humans. Scientists who wish to do experiments on humans or animals must have their research plans approved by a committee of fellow scientists. These rules are to help insure that human and animal subjects are treated properly.

SUMMARY

There are risks with everything we do. Taking proper precautions and using safe procedures can reduce these risks. Cooking, for example, can be dangerous. But you can cook safely by being careful and following safety procedures that reduce the danger. That's why people use potholders and keep pot handles pointed in toward the stove. When you conduct your science experiment, practice good safety procedures, too. Safety is no accident; plan for it.

Glossary

abstract—a brief summary of a research paper or report. It includes a brief description of the problem, hypothesis, procedures, results, and conclusions.

average (mean)—the most central or typical value in a set of quantitative data; the formula for calculating the mean is:

$$\text{mean} = \frac{\text{the sum of all the measurements (or counts)}}{\text{total number of measurements (or counts)}}.$$

axis—the horizontal or vertical line found at the bottom and left side of a graph; plural is axes.

back-to-back stem-and-leaf plots—a plot of two sets of data that are compared by having the leaves extend in opposite directions from the same stem.

bar graph—a pictorial display of a set of data using bars to indicate the value, amount, or size of the dependent variable for each level of the independent variable tested. Usually the taller the bar, the greater the value of the dependent variable.

bias—a statistical error that occurs when samples are drawn so that all members of the population do not have an equal chance of being included.

bibliography—a list of all books, papers, journal articles, and communications cited or used in the preparation of a report or scientific research paper.

box plot—an exploratory plot of a data set that displays the median and information about the range and distribution or variance of the data such as the minimum, maximum, and each quartile. Also called a box-and-whisker diagram.

call number—the set of numbers and letters used in a library to identify, catalog, and shelve each book.

central tendency—the value that is most typical of a set of data.

chi-square test—a test for qualitative data used to determine if differences between frequency distributions are statistically significant.

cluster—a subset of data within a data set, whose values are very close together.

conclusion—the last section of a report of an experiment; it states the purpose, major findings, hypothesis, a comparison of the findings of this experiment with other experiments and recommendations for further study.

constants—those factors in an experiment that are kept the same and not allowed to change or vary.

continuous data—measurements made using standard measurement scales that are divisible into partial units. Time and volume are examples of continuous data. When both the independent and dependent variables result in continuous data, the data can be graphed as *either* a line or a bar graph.

control—the part of an experiment that serves as a standard of comparison. A control is used to detect the effects of factors that should be kept constant, but which vary. The control may be a "no treatment" group or an "experimenter selected" control.

counts—data stating the number of items, for example, the number of bees attracted to sugar water or the number of people responding to a noise.

data—the bits of information (measurements, observations, or counts) gathered in an experiment; data takes a plural form verb, "the data *are;* the data *were.*"

data table—a chart to organize and display the data collected in an experiment.

degrees of freedom—the number of independent observations in a sample. In a sample of n members with a fixed mean, the degrees of freedom are (n – 1).

dependent variable (responding variable)—the factor or variable that may change as a result of changes purposely made in the independent variable.

derived quantity—information or values determined by calculations using collected data; examples are means, medians, modes, and ranges.

descriptive statistics (summary statistics)—statistics that describe for a set of data the most typical values and the variations that exists within the set.

discrete (discontinuous) data—data that exists in categories that are separate and do not overlap such as brands of products and kinds of papers. When displayed by a scale on a graph, the points between the defined categories do not have any meaning. Discrete data can be graphed as a bar graph but *not* a line graph.

error (in measurement)—the unavoidable chance errors that can result in increasing or decreasing the value of individual measurements.

experiment—a test of a hypothesis. It determines if purposely changing the independent variable does indeed change the dependent variable as predicted.

experimental design diagram—a graphic illustration of an experiment containing the components illustrated below.

Title:
Hypothesis:

IV:					independent variable
					levels of the IV tested and control
					number of repeated trials

DV: dependent variables
C: constants

experimenter selected control—the set of trials conducted for a single level of the independent variable that is selected by the experimenter to be the standard of comparison. For example, in an experiment to determine which brand of gasoline is best a "no treatment" control, using no gasoline in a test car does not make sense.

Four Question Strategy—an approach for generating a series of experiments from a topic, demonstration, or other prompt. Brainstorming responses to the four questions in the strategy results in many potential independent variables and dependent variables to select among. It also results in descriptions of ways to describe or measure the dependent variable and identify many constants.

frequency distribution—a summary or graph of the amount of variation (spread) within a set of qualitative data (observations); a frequency distribution states the number of items in each category, for example, 2 red, 10 pink, and 25 green tomatoes.

gap—data values within a data set for which there are no entries.

graph—a pictorial display of a set of data.

horizontal (X) axis—the line along the bottom of a graph on which the scale for the independent variable is placed.

hypothesis—a prediction of the relationship of an independent and dependent variable to be tested in an experiment; it predicts the effect that the changes purposely made in the independent variable will have on the dependent variable. Plural is hypotheses.

independent variable (manipulated variable)—the variable that is changed on purpose by the experimenter.

inferential statistics—the methods or mathematical procedures by which to determine whether the variations between sets of data are statistically significant or not. The statistics that indicate the likelihood that the variation occurred by chance or as a result of the treatments.

interquartile range—the difference between data values of the upper (Q_3) and lower (Q_1) quartiles, e.g., $Q_3 - Q_1$.

interval data—measurements made using a scale with equal intervals, but no absolute zero. For example °C.

intervals—the equal size values represented by the equal spaces marked along the axis of a graph, or the spaces between units of a measuring device.

introduction—a paragraph at the beginning of a report of an experiment that states the reason why the experiment was done; what was expected to be learned (purpose) by doing it, and the hypothesis tested.

J-shaped distribution—a distribution of data in which the value of the data that occurs at the greatest frequency is at the extreme of one end of the data values, indicating a probable limit of values.

level of significance—the level of probability set by the experimenter for rejecting the null hypothesis. It is the level of probability that the experimental results were due to the treatment and not to chance.

levels of the independent variable—the specific values (kinds, sizes, or amounts) of the independent variable that are tested in an experiment.

line graph—a pictorial display of data that can be drawn when the data for both variables are continuous data. The line in a line graph shows the relationship between the independent and dependent variables.

line-of-best fit—a smooth line drawn so that the totals of the distances between the line and the points above and below it are equal (roughly half the data points are above and half are below the line). The line that can be either a straight line or a smooth curve shows the relationship between the independent and dependent variable.

lower extreme—the minimum data value in a data set.

lower quartile (Q_1)—the number (data value) below which 25% of the values in a data set fall.

manipulated variable—see *independent variable*.

mean—see *average*.

measurements—data collected using a measuring instrument with a standard scale.

median—the central value in a set of data ranked from highest to lowest. Half the data are above it and half are below.

middle quartile (Q_2)—the median; the data value below and above which 50% of the values in a data set fall.

mode—the most typical or central value of a set of qualitative data. It is the value that occurs most often in the set.

negative association—the inverse relationship that exists when increasing the independent variable results in a decrease of the dependent variable.

nominal data—data for a series of discrete categories for which there is not a basis for rank ordering, for example gender or hair color.

no treatment control—a control that receives none of the independent variable, for example in an experiment testing the effect of varying the amount of a fertilizer on plant growth, a no treatment control would be a set of plants that receives no fertilizer.

note card—a card used to record notes about the information found in a reference source or interview. Reference documentation is also put on each note card.

normal (bell shaped) distribution—a distribution of data in which the value of the mean is the most frequent data value and at each distance above and below the mean there is an equal frequency of data values.

null hypothesis—a hypothesis based on the assumption that two samples are from the same population and therefore have identical means or means which differ no more than would be expected by chance.

observations—data that are descriptions of qualities such as shape, color, and gender.

ordered stem-and-leaf plot—a stem-and-leaf plot in which the data has been entered in rank order usually from the smallest value to the largest.

ordinal data—data collected for categories that can be rank ordered.

outliers—data values $< Q_1 - 1.5(Q_3 - Q_1)$ or data values $> Q_3 + 1.5(Q_3 - Q_1)$. These data values are considered to be unreasonable.

population—all members (persons or things) of a specific group that share a set of common characteristics.

positive association—the relationship that exists when increasing the independent variable results in an increase in the dependent variable.

probability (p) of error—the level of chance that the researcher erred in rejecting the null hypothesis.

procedure—a sequence of precisely stated steps that describe how an experiment was done, including the materials and equipment used.

qualitative data—verbal descriptions or information gathered using scales without equal intervals or zero points. Such scales are non-standard scales.

quantitative data—information (data) gathered from counts or measurements using scales having equal sized intervals and a zero value. Such scales are standard scales.

range—a measure of how a set of measurements or count data is spread out. It is calculated by subtracting the minimum value from the maximum value.

ratio data—quantitative data collected using a scale with equal intervals and an absolute zero.

reference documentation—the information needed to identify each source used in doing and reporting an experiment; this includes the author's name, the titles of the articles and or books, newspapers, or journals, as well as the date, city and state of publication, and the publisher; or, in the case of an interview, the name of the person interviewed and the time, date, and place of the interview.

reference style manual—a book that states rules for writing the reference information for the books, interviews, encyclopedias, magazines, journals, and newspaper articles used in doing and reporting an experiment.

repeated trials—the number of times that a level of the independent variable is tested in an experiment or the number of objects or organisms tested at each level of the independent variable.

responding variable—see *dependent variable*.

results—a section of the report of an experiment that includes the data tables, graphs, and sentences that summarize any trends found in the data.

sample—the specific set of individuals, selected from a population, to be the subjects in an experiment.

sample bias—a sampling error that occurs when samples are drawn so that all members of a population do not have an equal chance of being included.

scale—a series of equal intervals and values placed on each axis of a graph.

skewed data sets—data sets in which there are more data values above or below the mean or median.

standard deviation—the square root of the variance. It is a measure of how closely the individual data points cluster around the mean.

standard error of the mean—the statistic that indicates the extent to which sample means from the same population are expected to differ.

standard scale—a scale of measurement that has both a defined zero point and equal intervals. Examples are distance scales in inches, centimeters; temperature scales in degrees Celsius or Fahrenheit.

statistic—a number used to describe or analyze a set of data.

statistics—the branch of mathematics involved in collecting, analyzing, and interpreting data.

statistical significance—the criterion for the decision that the results of an experiment did not happen by chance but were the result of the treatment.

stem-and-leaf plot—an exploratory means of plotting data based on place value. The greatest place value, such as the tens or hundreds, is the stem, and the smaller place value, such as the ones, is the leaf. This plot shows the distribution of the raw data.

***t* test**—an inferential statistical test that is used to determine whether significant differences exist between the means of two samples.

title—a statement describing an experiment or data table. Titles are often written in the form, "The Effect of Changes in the Independent Variable on the Dependent Variable." In a title all words containing four or more letters are capitalized.

trend—the general direction or pattern of the data; it is usually illustrated on a graph as a line-of-best fit.

types of data—the kinds of information collected in an experiment; typical types are measurements, counts, or observations.

U-shaped distribution—a distribution in which two values of the data occur in higher frequencies than the other data values.

unordered stem-and-leaf plot—a stem and leaf plot in which the data is entered in an unranked order.

upper quartile (Q_3)—the number (data value) below which 75% of the values in a data set fall.

upper extreme—the minimum data value in a data set.

value—the size, amount, or extent of a property described by a piece of data, count, or observation.

variable—things or factors that can be assigned or take on different values in an experiment.

variance—the square of the standard deviation or the average squared distance from the mean.

variation—statistics that describe how spread out the values in a set of data are.

vertical (Y) axis—the line drawn on the left side of a graph on which the scale for the dependent variable is placed.

8006